Peter Henry Emerson, T. A Cotton

Birds, beasts and fishes of the Norfolk broadland

Peter Henry Emerson, T. A Cotton

Birds, beasts and fishes of the Norfolk broadland

ISBN/EAN: 9783743383791

Manufactured in Europe, USA, Canada, Australia, Japa

Cover: Foto ©berggeist007 / pixelio.de

Manufactured and distributed by brebook publishing software (www.brebook.com)

Peter Henry Emerson, T. A Cotton

Birds, beasts and fishes of the Norfolk broadland

BIRDS, BEASTS, AND FISHES

OF

THE NORFOLK BROADLAND

BIRDS BEASTS AND FISHES
OF THE NORFOLK
BROADLAND

BY

P. H. EMERSON

B.A., M.B. (CANTAB), M.R.C.S. ENG., A.K.C.

FELLOW OF THE ROYAL HORTICULTURAL, METEOROLOGICAL, AND PHOTOGRAPHIC
SOCIETIES, AND MEMBER OF THE SOCIETY OF ARTS

AUTHOR OF

"A SON OF THE FENS," "WILD LIFE ON A TIDAL WATER," "ON ENGLISH LAGOONS,"
"TALES FROM WELSH WALES," "SIGNOR LIPPO," ETC.

Illustrated with Sixty-eight Photographs by T. A. Cotton

LONDON

DAVID NUTT, 270-71 STRAND

1895

Printed by BALLANTYNE, HANSON & CO.
At the Ballantyne Press

PREFATORY NOTE TO THE SECOND ISSUE

THE present issue remains unaltered, save that some passages in Chapter LXIX. of Part I., and XVIII. of Part II., which were held, whether rightly or wrongly need not here be discussed, to make the book less suitable for general reading, have been left out or changed, and that the Preface to the First Edition has been omitted.

As the author's statement that many of the illustrations were taken from life has been implicitly or explicitly denied by some of his reviewers, it is here emphatically reaffirmed.

The author's terminology and classification are not those of official "natural histories," and some reviewers have assumed that when he differs from the "authorities" he must be in the wrong. He thinks it well to state that he maintains the accuracy of every fact recorded in the following pages.

He has set down nothing save on the warrant of his own observation extending over many years, or on the authority of fowlers, ratcatchers, fishermen, gamekeepers, and the like, whose work brings them into daily contact with the birds, beasts, and fishes of the Norfolk Broadland.

ERRATA

age 23, line 32, *for* "gauge" *read* "gage."
" 37, " 4, *for* "soft" *read* "loud."
" 37, " 21, *for* "April" *read* "summer."
" 38, " 3, *for* "as the" *read* "as that of the."
" 38, " 22, *for* "rond" *read* "road."
" 57, " 14, *for* "mice-like" *read* "mouse-like."
" 71, " 6, *for* "pied" *read* "white."
" 71, " 21, *for* "parnassian grass" *read* "water crowsfoot."
" 72, " 20, *for* "florets" *read* "leaflets."
" 99, " 12, *for* "brown" *read* "hard."
" 106, " 15, *for* "water-lilies" *read* "water-violets."
" 112, " 35, *for* "for they" *read* "for, to all appearance, they."
" 145, " 23, *for* "themselves" *read* "themselves entirely."
" 145, " 25, *for* "after the manner of a pigeon" *read* "an opera-
          tion which, at a distance, has the appearance of a
          pigeon feeding her young."
" 151, " 1, *for* "swallow family" *read* "popularly called swallow
          family."
" 166, " 12, *for* "heteræ" *read* "hetairæ."
" 208, " 4, *for* "ruddy" *read* "common."
" 210, " 23, *for* "Everett" *read* "Everitt;" *for* "Oulton" *read*
          "Oulton Broad."
" 219, " 14, *for* "diving" *read* "dive."
" 240, " 31, *for* "budding" *read* "leafy."
" 243, " 9, *for* "wakes" *read* "walls."
" 250, " 32, *for* "shout" *read* "whistle."
" 255, " 13, *for* "on" *read* "over."
" 277, " 18, *for* "till he is about to fly up" *read* "look."
" 277, " 22, *for* "their" *read* "three."
" 278, " 22, *for* "but" *read* "they."
" 279, " 26, *for* "down" *read* "up."
" 285, " 17, *for* "vocal" *read* "laryngeal."
" 292, " 3, *for* "simple-like" *read* "snipe-like."
" 292, Plate, *for* "black-tailed godwits" *read* "bar-tailed godwits."
" 309, line 6, *for* "pile" *read* "plod."
" 309, " 26, *for* "night-jar" *read* "rook."
" 313, " 19, *for* "stomach" *read* "crop."
" 320, " 4, *for* "turf-decks" *read* "turf-deeks."
" 320, Plate, *for* "sea-gull" *read* "sea-birds."
" 332, last line, *for* "deck" *read* "deek."

CONTENTS

PART I.—BIRDS

PART II.—BEASTS AND FISHES

(MAMMALS, FISH, AND REPTILES)

LIST OF ILLUSTRATIONS

PART I.—BIRDS

xiii

PART II.—BEASTS AND FISHES

(MAMMALS, FISH, AND REPTILES)

" They do not sweat and whine about their condition,
  They do not lie awake in the dark and weep for their sins,
  They do not make me sick discussing their duty to God,
  Not one is dissatisfied, not one is demented
  With the mania of owning things ;
  Not one kneels to another, nor to his kind that lived thousands
      of years ago ;
  Not one is respectable or unhappy over the whole earth."

—WALT WHITMAN.

PART I.—BIRDS.

GATHERING SEA-BIRDS' EGGS (YORKSHIRE).

A

" Papillon, tu es volage !
    Tu ressemble à mon amant :
    L'amour est un badinage,
    L'amour est un passe-temps :
        Quand j'ai mon amant
        J'ai le cœur content."
                —*Chansons Populaires.*

EGG-GATHERERS.

BIRDS, BEASTS, AND FISHES

OF THE

NORFOLK BROADLAND

CHAPTER I

THE SPECKLED THRUSHES

THE speckled birds of the coppice and field are of varying
degrees of distinction and attractiveness.

The familiar *Song Thrush,* or " *Mavish,*" as the fenmen
call him, is our dearest songster. He is, moreover, the most
delicate of the speckled tribe, although he is the earliest
riser of all the birds, singing of a morning before the lark.
When the grey mists of February soften the outlines of
the farm-buildings so that they hang in the air like castles,
and the lone trees rise from the upland like weird ghosts,
the mavis begins to sing his beautiful love-song with its
series of triple notes ending abruptly with "pretty boy."
He delights to sit of a morning on a bare sprig of thorn-tree
overhanging a holl, and pour forth his song ere he goes a-
hunting for worms, startling them from the soft ground, and
killing them (when caught) as a heron kills an eel. After
eating but half of his slippery quarry mayhap, he goes a-
dodman gathering, carrying his prey to some stone-anvil by
the roadside to crack the husk that guards the succulent
flesh. And if dodman-meat taste as sweetly to him as the

3

Escargot de Bourgogne does to us, well may he seek it far and wide.

At this season, too, he delights in fighting his rivals, and you may see them rolling over and over in battle on the marsh-walls or by the roadside. At eventide, too, when the grey mists are closing over the lush flatland, you may hear the cock-birds singing around you until darkness hides the landscape, when he suddenly stops his song, being followed by the screech of the owl.

When the bitter winds of March blow dryly over the land, the mavises begin to build their mud-fashioned cradles, laying their sky-blue speckled eggs long ere the last chill snows have fallen. Indeed, you may see some snowy day a hen sitting in the heart of a bare thorn-tree, her head resting on the edge of the nest, her feathers shot, her tail contracted, looking like a serpent, her lustrous eyes staring at you through the cold snow-crystals as they fall, powdering the bleak marshlands till the rivers look like leaden skeins threading a white world. At any time up to harvest you may find the mavis sitting, for if robbed she will build again and again; and all through the bright-flowered spring and warm-scented summer you may come upon the young birds, bright in colour, flying chiefly through sallow coverts, gorse coppices, and hedgerows that lead to the uplands, or mayhap skulking in clumps of undergrowth of gorse on the river-walls. Should the lone marshman possess a black-heart cherry-tree in his garden, or a patch of strawberries, the thrush knows it, and you will see him running briskly over the ground under the lowest branches, or hopping lightly upon the light sprigs bent down with fruit, or else pecking daintily at a luscious British queen strawberry in the patch. Nor is he averse from currants and gooseberries, and the marshman, who knows this, keeps watch over his scanty unpruned bushes with old muzzle-loader, as many a mavis and black-bird knows full well; for often the child-like screams of the maimed birds resound far and near. A cinder-heap, too, is

dear to the mavis; but that he is free of. And when the
hard toil of feeding the young is over-past—for the fledglings do not leave the nest till they can fly—the families
form flocks, and go a-foraging for food amongst the fruit
crops or along the upland hedgerows. Not much singing is
done after this flocking—the cares of life press too heavily
upon them. And when the dikes are ice and the marshes
slabs of snow, the mavises flock to the hawthorns, scarlet
with berries, whence their speckled breasts gleam in the
autumn sunshine, as they fight for the haws as parrots
wrangle over wild oranges.

Mayhap at this season a keen-set sparrow-hawk swoops
down upon one here and there, and the biting cold kills many
more; and though the flocks are increased by migrants from
over the seas, still, when winter is gone, and the fighting
season returns on foggy wing, their numbers are sadly
thinned. Nevertheless, they are not cursed (or blessed) with
memories; and when the love impulse guides them to it,
again they burst into joyful song, and once more spring,
the season of the Lent lily, the primrose, the white violet,
and the pale flowers of the coppice is upon us, and our
hearts beat with gladness.

The Missel Thrush, or " Fulfar," or " Yellow Fulfar," as
he is called on the marshes, is a better fighter but a poorer
minstrel than the mavis; his temper is too hasty to allow
him to be a great artist. Yet 'tis a brave bird, fighting for
love unto death; rolling over and over, struggling until one
slinks off into a corner to die, and leaves the victor to sing
heartily through the storms that sweep across the marshland. Their fights take place in February, just before the
birds pair; for the old Fulfar is one of the earliest birds to
begin his wattle and gutter-putty nest in the crotch of an oak,
wherein the four greenish eggs with rusty spots are laid.

At this season they are " artful old seeds," as the farmers
say; for do you approach their half-built cradles, and they
will sneak off at once, and never go near the nest until

you be gone about your business.   But should you discover the finished nest and go to rob it, the fiery couple will
attack you right bravely, making a noise that puts you to
very shame, and flying straight at your hat, as if to lift it
from your head.   You or the thieving cuckoo, he cares not
which—you or any other foe, he will fight and drive from
his precious eggs, with all his bravery—an he can.   Whilst
his hen is sitting closely in some hollow tree or marsh-oak,
he will be sneaking about the trees and bushes, driving off
any strange bird, for he is of a dog-in-the-manger disposition.
And should you rob him, his wife will go laying again and
again, ay! unto ten clutches, so anxious is he to propagate
his kind and seize the world to himself.   And so closely does
the hen stick to her eggs, that you may sometimes take her
from the nest with your hand; and once having found her
cradle, you know where to find it year after year (an she be
not robbed), for spring after spring she returns to the same
tree, and to the same crotch of the tree, to build.

The young do not leave their nest till they too can fly,
and then they get to know the country round their cradle
in their small excursions for worms, snails, and slugs, on
which they feed.   The marsh cottager, too, knows the old
birds, for they dearly love a cherry, or even the buds of
plums, bullaces, and red-currants; but his brave song does
not repay the stolen buds and fruits.

In autumn you see them keep to themselves, family by
family, for they are unsocial and independent birds.   Indeed, you rarely see more than four or five together in a
bunch on the land, or walking clumsily amongst the turnips
or along the holls.   But they love best to keep about the
hawthorn trees, gay with red berries.

In the hardest weather, when the rivers are frozen, you
may see them in large flocks in company with the fieldfares, but they are not great friends, for you may see fighting
going on at one end of a row of bright-berried thorn-trees
in the same field, whilst at the other end the birds are

greedily swallowing the berries whole, throwing the stones up as a second thought, ere they fly across the low red disc of the setting sun to roost in a friendly holly-tree; for a hardy bird is the missel-thrush, rarely succumbing to the frigid winters of the marshlands: he is too brave to die of mere cold; like the hero he is, he prefers to die fighting.

The Fieldfare, or "Dow Fulfar," as the fenman has named him, is merely a visitor, coming over in large flocks in autumn after the larks, and with the redwings, to frequent the marshes. According to the fenman's simple calendar, the old "dow fulfar" is to be seen directly after the last morfra has carried home the yellow grain to the stackyard; and the fenman is right. He is a punctual bird, as I have observed season after season.

He is a lover of open spaces, merely taking to trees when the heavy gales and hard frosts drive him there for food; and even there he perishes in numbers, his speckled breast dotting the icy ground beneath the bare coppices more frequently than any other bird; for he is a tender creature, and lacking in character—a bird doomed to extinction, I surmise.

When an Arctic winter freezes the rivers, and he draws up to the land a bit, you may see him looking about for food amongst "newlays" and by marsh-roads. Then, too, you may recognise his note, differing as it does from the rest of the tribe, his winter voice being a harsh "*sack! sack!*" whilst the redwing makes a "weeping" noise. He is much bigger, too, than the redwing, and though hardly as interesting or familiar, yet he loves the wild wastes. He has, when delayed by a cold spring-tide, dropped his eggs in England, but never in the Broad district that I know of, though I have seen fieldfares there as late as May, sitting on a pasture by the roadside.

Lastly, the smallest of the family is the little Redwing, or "*French Mavish*," as the men here call him. He, too, is a winter visitor, and at times mixes freely with the flocks of fieldfares, but more frequently he is to be seen about in small flocks feeding on the scarlet haws that shine against the snow; for he is hardier than the fieldfare, and tamer too. For when the milk freezes in the pail, you may hear his little chicken-like "weeping" voice near the bottoms of stacks, or upon newlays and ollunts close by the marsh farmhouse, whence they rise at times in hundreds, as the farmer goes forth to get a hare for his dinner. But in the hardest weather they too perish, and you may find their stark frozen corpses under the leafless hedges, unknown, unnamed, unburied, in a strange land, far away from the home of their birth, though the broadsmen tell you that in days gone by the French mavishes used to nest in the reed-beds.

CHAPTER II

THE BLACKBIRD

THE blackbird is the emblem of greediness and noisy vulgarity, of selfishness, and assertive pretentiousness, and his chief virtue is his pugnacity during the mating season.

When the dikes are ablaze with kingcups, the cock-bird sings his thrush-like song noisily from the low hedgerows until a rival cock appears, flying with a rush at the peaceful songster, who, often taken by surprise, flies forth from his perch, with many noisy calls, to the marsh, where he alights with raised tail to recover his balance. In a moment his rival is upon him; their heads are erect, their wings slightly spread out, and they dash at one another, fighting and darting, and shifting ground across twenty or thirty yards of marsh, when they will fly up into the air, fly at each other, alight again on the marsh, and go at it until one feels himself worsted, when he flies off, crying a harsh "*puk, puk, puk,*" seeking refuge in some friendly hedge, leaving the victor to court the plain hen-bird.

Together these twain build their artless nest of mud and horse-dung in the hedge, or behind some sheltering stump, where their useless eggs are laid, and the ugly brood reared.

As soon as the nestlings can fly the family seek the seclusion given by a neighbouring alder-tree—a tree near the water for preference. There they live, going early to bed and rising late, starting off to steal their breakfasts from the nearest garden, to rob the cottager of his cherries or the gardener of his figs. They lack no cunning in their thieving, for they will walk along the ground and jump to the lowest berries, catching the fruit in their bills and hanging on till the slender stem gives way—in fact, pulling them off by sheer weight.

After the breeding season the cock-bird stops his song —it was used for selfish ends, for he is now content; and when the family "can do for themselves" they separate, seeking the gardens dotted over the marshes or fringing the villages. Nor do they do much good to the gardens, as do the thrushes, starlings, and robins, but they merely gorge strawberries, cherries, currants, gooseberries, &c.; in many a garden, moreover, they seek out the choicest peaches, nectarines, and dearly love ripe figs—having the impudence to fly into the fig-tree on one side whilst you are actually getting fruit on the other. Peas, too, they delight in as much as does the sparrow. He is a bold bird the blackbird, but he weeps when his nest is robbed, wailing *chi-chi-chi*, though he does often fly boldly at the robber's head and face. In truth, his audacity knows no bounds. One of these pleasant thieves built his nest in an old English codlin-tree growing in the middle of my strawberry patch, which nest I was glad to rob, for I have never yet seen a blackbird do good in the garden.

Most insulting, too, is his vulgar cry as he flies away over the wall—a mingled vulgar cry of alarm and derision— a chortling that makes one glad when the winter cold drives him to the homely fruit of the hawthorn—a dish too good, however, for any but thrushes, who gratefully repay their pilferings of fruit by killing snails. So much do I detest him, that I have often wished an old peasant's love of black- bird pie was a common taste. This old man used to fre- quent the berried hedgerows with an old blunderbuss killing blackbirds, and so peculiar was this taste, that the keeper of those hedgerows began to suspect the blackbird-killer, and one day he followed him up a lone hedgerow. The old fellow went along calling "Pie, pie, pie!" until at last a fine cock blackbird flew out. "Fine cock! fine cock!" muttered the old man, and "poff!" he fired and killed his bird. When he put the cock into his pocket, the keeper's hand went with it, and drew forth six blackbirds for the old man's pie. The old fellow had found the proper use of the blackbird.

Nearly every winter evening I hear his vulgar call, as he flies at closing-in time to roost in the evergreens, leaving me longing to go south, far away from dreary mists and melancholy landscapes.

BLACKBIRD'S NEST IN A CLIFF (YORKSHIRE).

RING-OUZEL'S NEST.

CHAPTER III

THE RING-OUZEL

WHEN the hedgerows grow green with buds at the end of April, and the sealike marshes are green with springing oats, small flocks of ring-ouzels come over the gleaming sandhills from the sea and scatter over the fields.

You may know this bird from afar, for he is not so vulgarly "smart" as the blackbird, and is a lover of the open fields. The blackbird skulks in the hedgerow or beneath the pendant fruit crops, but not so the ring-ouzel; for, though a great lover of fruit, he leaves the hedgerow and seeks his food afield. You may see them in early spring on any of the marshes by the sea feeding upon worms; and should you disturb them, they will, blackbird-like, fly off to the hedgerow, where they occasionally build their blackbird-like nest and lay their starling-like eggs, scant of spots.

But the ring-ouzel is not common in the Broadland, and beyond a few flocks in the early spring, and scattered wanderers a little later when the ivy is in bloom, and a very few nesting couples, he soon leaves the marshlands for the uplands, and is not seen again until autumn with many-coloured wings settles on the land, when the ring-ouzel passes on his return journey bound across the sea, his plumage rather frayed and weather-beaten perhaps, but still the familiar ring-ouzel.

CHAPTER IV

WHEATEARS

As you walk along the warrens, where the yellow silver-weed blooms at the edge of the sand-dunes, you may in early spring come upon the first pair of wheatears as they flitter across the marram green hills and show their white rump-bands; for upon their arrival across the grey sea they frequent the warrens at first—it is warmer there, and they can feed upon the early flies and sand-bred insects. And this practice of theirs has earned them the nickname of "coney-suckers," their common name amongst the Broads-men, for they say these birds enter the "rabbit eyes" in the dunes and suck the milch-does.

One spring, whilst walking along the crest of the sandhills, amongst clumps of honeysuckle, juicy oak saplings, brakes of bramble, blackthorn, bracken, and flowering gorse—for such is the lean flora that flourishes amongst the marram's fibrous roots — I suddenly saw two birds fly up before me, and flying on, alight on the sand, standing very up-rightly, and looking very black and white against the sandy background. Sinking into the soft warm marram, I watched them feeding on insects, working like a robin, though more eagerly, and cheeping like a titlark. Every now and then they would pause, and, standing upright, look as elegant as any courtly lady. Then they flitted on to a spray of gorse, and so I followed them to the round steeple that stands without the sandy fortalice.

I knew them from the first for wheatears, but after close shadowing and careful observation, my glasses proved them

to be wholly unknown to me—for they were the first and
last of the kind that I ever saw in the open—but from
careful notes made on the spot as to their markings, I was
afterwards able to identify them as desert wheatears; the
black tail, throat, and buff head marked them for every one
as desert wheatears. These birds were not so plump as
the common wheatear, but more graceful, and stood up
more straightly. They walked differently too—more like a
robin.

Howbeit, as we returned over the dunes in the roar of
the fresh sparkling sea, noting the footprints of rat and
rabbit, and the dung and spoor of partridges, who dearly
love the sandhills, we saw some of the common wheatears on the beach, feeding upon insects; and as we walked
home across the marshes, now turning emerald, we passed
single wheatears standing by the marsh dikes or feeding
upon the marsh-walls, flying up safely out of reach, and
calling with a peculiar note, jerking their tails at the same
time.

Though never numerous, you may nearly any time during
the summer see a pair or two of common wheatears as you
walk over the marshlands or by the sea, where I have seen
them flying from stone to stone.

At that season, too, you may see a hawk chase them, but
never have I seen a hawk cut them down, nor yet have I
ever found a nest. One old fenman tells me he has found
their nests on the warren in a rabbit's hole, also by a dike in
an old rat-hole; but that was many years ago.

Nevertheless, I am inclined to think they nest in the Broadlands, one sees them about so often and so long, though I
must say I have never seen any very young birds.

They are rather mysterious birds, for although you see
them flicker across the landscape or start from a dike where
the bottom-fyers have recently been working, yet they do
not let one see much of their lives—not even the "coney-
sucking" with which they are accredited. In truth, the

wheatear is a bird of whom we know little in the marshlands, for he comes in spring, flits up by the summer
dikes as you pass, and goes away again in August, so that
the chief memory we have of him is his soft song and the
vision of his flickering presence.

COMMON WHEATEAR'S NEST AND EGGS.

CHAPTER V

THE WHINCHAT

WHEN the soft grey showers gleam across the marshlands in silvery April, little flocks of whinchats appear
amongst the dead thistles and gorsy clumps of verdure on
the flatland; a gay cock and two or three sober hens,
generally feeding together, hunting over the marshland for
flies and moths, their favourite food. Indeed, they follow
the marsh-mowers, with other birds, to snatch the moths
from the swathes, as those insects fly startled from the
falling marsh crops. The gaily painted little grasshoppers,
too, are a staple dish. You may lie upon the marsh, hidden
in tussocks of rush, and watch the lively birds flying from
a gorse spray to the ground to feed, and you will see them
flit back calling "*utick, utick ;*" or perhaps one will sit on
a spray just above the marsh, and sing a pretty sweet little
song.

Then something seizes their minds, and they flit off to
another gorse bush, and again the same ceremonials are
gone through. By the beginning of June they are paired,
and build their hedge-sparrow-like nest of grass, often lined
with the flaxen down of the gladen spindle, in rush bushes,
brambles, or gorse plants that clump together on the marshes
in prickly islets. They prefer these islets to be by the dike-
side, for they delight in the presence of water, and there
by the dikes, gay with flowers, you may hear their short
calling whistles. And should you perchance see them
whistling, you will see their tails jerked up every time they
speak. Or at times you see them sitting upon a fork-

shaft left standing on the levels; they delight to perch on a fork-shaft as much as does the cuckoo. And when the large-sized eggs—recalling a reed-bird's—are hatched, and the fledglings can fly, you again see the little families flitting over the marshland feeding on the gaily enamelled grass-hoppers and soft-winged moths. And in winter, though most of these birds go across the seas, still a very few remain and are to be seen on the uplands. One winter I saw a cock-bird amongst the yellow reeds by the Waveney; but it is a very rare bird in the winter landscape. Still it is seen, a fact, I believe, disputed by some naturalists. On the whole, it is a quiet, simple, unobtrusive little bird, without much character; a harmless, pleasant little creature. More-over, the cock, in full plumage, is a welcome spot of colour in the sober-hued landscape of early spring.

CHAPTER VI

THE STONECHAT

THE Stonechat, too, loves the Broadland, and his brighter colouring makes him more desirable than the whinchat. He might have been appropriately called the wheatear, for his song, a harsh, abrupt *wheat-ear, wheat-ear*, can be heard as he flits from fern to bramble-bush in the early spring-time; for he is not a lover of marshland when the snow-blast and hail-squall sweeps over the land, rustling the creaking reed-stubbles. Then he retires to the uplands, where hedge-row and farm-shed give him shelter when the rime-crystals deck the dead thistles with gems.

But in early spring he comes down to the marshes, and you may see him flying up into the air, catching insects, returning to his station on a thistle-stalk, whence he drops down to the marsh-sedges for some dainty morsel, flitting back again to his look-out perch.

And when the flowers are glowing over the grey marsh-lands, these little birds build in the side of a marsh-wall, placing their nests in the grass, or they may choose some marsh-bottom. Their cradle is built of marsh-grasses, hair, roots, and feathers, wherein six eggs—greener than those of his relative the whinchat—lie snugly. The parents never leave the nest far; should you approach, they keep "chatting" and flying restlessly about the nest, alighting on bramble-sprays, prickly gorse, or sallow-stoles.

Later, when the young are hatched, you may see them following the mowers, catching moths and grasshoppers and flies, as the coarse swathes fall before the sturdy marshmen's strokes.

In autumn the marshlands scarce know the stonechat, for he seeks his dinner of seeds upon the uplands.

Like the whinchat, he is not too common, and being a little, shy, colourless character, there is but little to record about his life. He and his cousin the whinchat are playful little children of the marshland—children who, during the breeding season, take life more seriously, but still remain throughout children, who play at life, offending none. Long may they live to enjoy themselves!

CHAPTER VII

THE FIRETAIL

THE Redstart is rare in the Broadland, but the cock is a beautiful bird when he flashes early in April across the tender green of a newly-leaved thorn-hedge beneath a blue sky. He is very shy and very active, and all you may see in his passage is a patch of red and black chequering the background, or hopping this way and that from some spray to the ground and back, producing on the mind the effect of a red and black toy paper-mill. But his gayness leaves a bright impression with you, for you cannot keep him in sight long; directly he perceives you are following, up he whisks with a sharp turn across the hedge, and 'tis useless to follow. I have never found his nest in Norfolk, but old nest-finders tell me that he generally selects a hole in a tree — a pollarded willow for preference—and that he returns year after year to the same nesting-place, if not disturbed. But very few must build in the Broadland, for I have seen scarce one during the summer, and not until after harvest, when the marshes grow beautiful with the chameleon-like beauty of dying brilliancy, does the little firetail flicker across your path again, chasing the insects that are daily growing scarcer, as he knows full well; for in September he leaves us, bound on the old trail, "the trail that is always new," leaving us a gay memory of whistling red and black.

CHAPTER VIII

COCK-ROBIN

"Bob," as the Broadsmen familiarly call this pert, boy-like bird, is the most respected bird in the district; for does not the local rockstaff say, " Ef you rob Bob of his eaggs, you're sure to break your arm;" therefore Bob's eggs require no Wild Birds' Protection Act in Broadland. Yet these red-breasted birds do not multiply exceedingly. They are such boys to fight—for their decrease is not due to bird-hunting cats; for though Pussy will kill Bob at times, as she will the young swallows, mavises, and starlings, for mere sport, yet Pussy despises Bob's flesh. Not so, however, with larks, sparrows, and blackbirds: Puss eats them and licks her whiskers afterwards.

In March, when the black earth is bright with yellow crocuses, or the porcelain-blue bells of the hyacinth, or later with crimson and gold tulip cups, Bob's melancholy song is to be heard in the hedgerows and bare coppices. Bob and another fighting fiercely for the lady, waging fierce battle, with loosely hanging wings and standing feathers. The rivals fly at each other like young bucks, until one retires from the contest, or is left dead amid the spring flowers, whilst the perky little victor goes off with his mistress to his station—chosen the previous autumn—and then the simple cradle of moss, leaves, feathers, or horse-hair is built in some hole in a bank. When the familiar little eggs are laid, the hen begins her duties, sitting very closely on her eggs, whilst the pert little husband goes in search of black worms, wherewith he feeds her upon the nest.

And should any urchin, regardless of the warning rock-staff, put her off her nest and rob an egg or two, she will not desert her home, but return to his leavings with a contented and philosophic mind, warming them into young fledglings, whom Bob will feed with wire-worms. He is busy then, in no mistake, carrying and giving the loveless wire-worms to his brood, himself eating any worm or " sow " (wood-louse) that he finds. Sometimes he finds a young cuckoo in his nest instead of his progeny; then he feeds him as carefully as he would have done his own children. And all summer Bob is breeding, for one family a year is insufficient, two or three broods being regularly turned off.

And when the autumn air is yellow with falling leaves, Bob begins, perhaps, to regret his uxoriousness, for his quickly growing, pert little sons are already as big as himself and lustier, and—oh! that it should be written—ungrateful; for with the fall of the yellow corn before the dipping harvesters the duels between father-robin and son-robin begin, and all along the embrowning hedgerows you may come across these fights, where fierce and unnatural war of upstart children against careful parents is being waged, and many an aged parent is left dead upon the battlefield, the homing harvesters at nightfall muttering their funeral sermons as they kick the dead bodies aside into the holls. And many an old Broadsman whom I have closely questioned assures me he has never found a young bird dead after these battles. One of these men, who in winter is a professional gunner, assures me he can find a score of dead birds any year he likes during the harvest season, all old birds. Nor is this little fighter content with battling against his own flesh and blood, but he must needs throw the gauge to the strong-beaked house-sparrow, and at times Bob worsts him.

When the fields are white and the misty hedgerows hang above the marshlands, Bob draws near to the fenman's cottage, getting scraps at the pig trough, seeds from the ricks, and eke crumbs from the fenman's table ; for often the kindly man leaves his door open, and Bob walks right in, " onconsarned as a passenger," and eats between his quiet host's crotch-boots as carelessly as a starling wanders under a horse's belly.

And though Bob's song be rare in winter, still you may occasionally hear it, and the fenman says with a smile, " Thar's old Bob a calling on you ter feed him." And feed this pretty ungrateful young generation of " Bobs " you will, for the little creature has such charming manners, and is so confiding—quite one of the mildest of parricides, yet selfish withal. Nothwithstanding everything, like the boy highwayman, Bob will always be a hero in popular sentiment. As for me, I like his pluck. His breast is a bright patch of colour in the landscape, moreover, and his song is not repulsive, if not very attractive. The sentiment aroused by him, too, is one of tolerant amusement. A desirable little fellow on the whole is Bob.

MUD COTTAGE, NORFOLK.

CHAPTER IX

THE NIGHTINGALE

WHEN the coppice-trees of the Broadland have broken into
bud, and young leaves look large and bland by the still
waters of the lagoons, the Nightingales that live thereabouts
arrive from over the sea, and mayhap as you are sailing
some evening over the silver mere, within a mile of the
sacred grove, their sweet *jug, jug*, breaks upon your ear,
as you glide noiselessly through the still waters, your sail
clearly reflected in the stiller water below; and with your
ears full of his sweet music—for a rival soon answers the
proud challenge—your eyes turn dreamily to the distant
wood, looming fresh and green over the flatland, and you
listen, soothed by the soughing of the breeze in your ear,
through which soft æolian music the distant voices of the
proud cocks steal like some far-away dream of music, reminding you the nightingales are back again to their old
haunts, to challenge and pair and build their coracle-like
nests of oak-leaves, in which to float their fledglings upon
the grassy marshland sea.

And next day, mayhap, you steal into the wood, and hear
first the regular strokes of an old man riving broaches, who,
apparently heedless of the music, splits up his wands for
the coming thatching season. But he has not been inatten-
tive, and as you draw near he sees you are absorbed in the
bird's music, and says carelessly, " You see that's a different
bahd a-singing t'mornin; he hev a different note; one bird
sing of a night and t'other sing of a daytime. I heard 'em
for the first time last night." And the spell is strong upon
you, and you listen and recognise the truth of the old man's

observation; but you answer never a word, the homely-looking little minstrel's song has sealed your lips: you are in the presence of art, of a joyous outburst of love, of a pæan of battle, such as one used to listen to day after day and night after night from rooms overlooking the old gardens of Downing College, where one used to recall the songs of the singers of Greece, accompanied by the nightingale's songs, which, like flowers, are "always old, yet always new."

Later in the season you are in the same wood again listening, for these birds are very rare in the Broadland. And as the song stops another bird begins to croak, and the old broach-river rises slowly, and strolling out with his black dog, as he adjusts his old moleskin cap, he leads you silently to an old tree-stump and points out a nest of young nightingales; but then you are not discomfited, for you know that pure love-song will last for six joyous weeks in the prime of the year, when all the world is young again, and every lad has his lass, and the world lives poetry, as it did in the golden age of Hellas.

NIGHTINGALE'S NEST AND EGGS.

CHAPTER X

THE WHITETHROATS

THE Greater * Whitethroat, or " Hay-jack," as he is locally called, is by no means uncommon in the gardens and hedgerows near the Broads.

When the lanes are white with May and the cherries are in bloom, the whitethroat's short, cheery song, half land-bunting, half-chaffinch, may be heard for the first time, for he comes from across the mysterious sea, when the blue spring air is dappled with moths' wings and traceried by the flight-lines of insects. For he is an insect-feeder; and when the fenman hears the hay-jack's song, he knows well the cheery sedge-warbler will soon follow.

Nor is he a laggard in love, for within a fortnight of his arrival his brief courtship is over, and the happy pair may be seen flying along the thorn-hedges building, their delicately tinted bodies flitting in and out, the cock being the architect, carrying the delicate materials—dead whip-tongue, grass, and roots—briskly making six or seven journeys in ten minutes, his white throat and buff coat shining softly between the green leaves of the hawthorn as he disappears with a mouthful of fine grass or horse-hair; for he takes but little pains with his cradle—indeed, you may often see through it, so slightly is it built. Nor does he confine himself to a thorn-tree, though it is a favourite site. Still you may see his nest in a bank where the violets bloom, in a tangled growth of brambles and dog-rose, on the border of

---

* Though I have seen a few pairs of the lesser whitethroat in this district I believe them to be rare.

a dry ditch where the blue speedwell flourishes, in a little clump of lush grass, or in a prickly bed of nettles, or even on a piece of waste land.

And when the eggs are laid and the hen sits closely, you may see the cock, of delicate coat, fly joyously up into the air for a dozen yards, returning to the exact bramble spray whence he started—a mere ecstatic expression of joy that all is thriving so prosperously in the little family circle. And when the young are hatched—those weak, timid fledglings that recall the sedge-warbler's young in habit—the happy pair are both busy catching flies and moths near the beloved children, for they never leave their nest far.

And as soon as the timid chicks have a few feathers—not a quarter the number they should have—these young things will be leaving the nest (again resembling the sedge-warbler in habit), though they are unable to fly across a good mill outlet: they think, perhaps, that they can run sufficiently well to escape an enemy. But they are not alert enough for that, and the hunting urchin soon "muddles them out," as he expresses it, or makes them lose their little heads, both metaphorically and often, I fear, in reality, for he takes them home to his father's ferrets—those omnivorous bird-eaters.

But when the cherries are ripe the parents know, for the "hay-jack" dearly loves a cherry, thrusting his delicate bill into the most luscious cheek of a Frogmore *Bigarreau* with the relish of a connoisseur. At other times you may catch his young (four or five in number) sitting closely together upon a bramble, forming a beautiful little decorative group against the blue, and if you drop behind the nearest green screen and watch, you will see the mother, mayhap, come with a full crop and alight on a spray opposite, and, 'mid many cheepings and flutterings of wings, the timid little songsters are fed *in turn*. And a pretty little picture it is, till some boisterous boy, perchance, as sometimes happens, knocks three of the timid weaklings over with a smooth stone, so closely are they seated at their little family meal.

But you are consoled when you think that they will rear two more broods that season, if not robbed **by** man nor beast —three families reared to leave with him when the September gales **blow** boisterously over the seas.

A delicate, timid little songster is the " hay-jack," but dear, for all that, with his softly-tinted plumage, delicate as the silvery petals of a rose, and his smoke-coloured crest, chaste as the bloom upon a peach ; and, in return for his sweet short song, he is most welcome to all the cherries he can peck in my garden, nor do I think the needy fenman would grudge him a white-heart.

CHAPTER XI

BLACKCAPS

MAY has come in with green and gold; the river is em-
broidered with flowering chate, and the sallow plantings are
green and cool with tender leaves and musical with flies—as
the blackcaps seem to know, for there they are flitting to
and fro through the willow wands, their little bills cracking
sharply together as they catch the midges in their flight from
stole to stole. Hour after hour they feed in the little willow
grove, the cocks preferring the tops of the pliant wands, but
every now and then dropping to the ground, to be followed
by their sober-hued little mates. They seem never to catch
enough flies to make up for their exhausting journey across
the blue sea—for the weather is fine and clear.

You must look well at them and listen to the sweet, little
song, for you will rarely hear them in the Broad district;
and if you find their nest thereabouts, consider yourself
lucky, for I have never yet seen their eggs in that watery
land; the birds themselves seem rare enough.

CHAPTER XII

GARDEN WARBLER

WHEN you walk along the green hedgerows, bright with new-born leaves — hedges running on either side of the grassy, springy lokes that lead from the marshes to the upland—you may often hear this warbler's sweet song and shriller chuckling call, but seldom will you see him—at least I have rarely been able to catch sight of him, and when I first made his acquaintance it was in an orchard in which grew some alders down by the water. I stood hidden in the alders, for I heard a bird then unknown to me, and watched, and by good luck he came into open view, hopping round the stems, stopping to sing every now and then, his little throat swelling forth with song; but I have always known him since, though he is very shy, flitting off directly he gets sight of you, whether he be high up in a thorn bush or in a garden catching caterpillars or stealing figs; for he is the first bird to find out that my topmost brown Turkey figs are ripening; indeed, he tells me they are ripe, for he is sure to have the first peck into the purple-brown skin, discovering the pink flesh; then you may watch him, for his love for figs overcomes his shyness. I have, in a Kentish garden, stood beneath the leathery leaves with my glasses upon him for many minutes, a matter I could never do in Norfolk. When disturbed, he merely flew up to a large roomy walnut-tree and watched till I had gone; but he is no fig-robber like the blackbird; after all, the garden warbler filches but little, and that generally over-ripe fruit.

I have never found its nest in Norfolk, nor, in truth, seen it there. Indeed, the bird is unknown to most of the fenmen, as is the blackcap, which, so far as my experience goes, is very rare in the Broadland. The garden warbler loves the lanes of Broadland as well as the cottagers' gardens, and you may see him fly across the road, or fly out of the light green hawthorn, going down the loke-cutting ahead of you, but he soon clears off to a hedgerow running up a field away from the road: he is shy, very shy, except when over a fig —that is his grand passion. If I did not know his song, I should have thought him rare indeed ; but he is commoner in the Broadland than is generally supposed.

CHAPTER XIII

THE HERRING-SPINK *

As the North Sea fishermen call that mighty soul in that little body—the Golden-Crested Wren—is rather rare about the Broad district, though numbers may be seen on the sea-coast nearly every autumn. But at times you see them farther inland where the sluggish water sleeps on the lazy weed.

One April morning I sailed over the sleepy tide between walls of fresh young gladen that shimmered with chameleon-like colours as the soft winds blew it this way and that—now showing dark as the green ribbons faced you, now shimmering as they waved idly from side to side, anon glancing and showing green when the wind filled your sail from behind and tugged at the sheet asking for more, and lastly dyed with blue along the edges—a pale blue reft from the azure overhead—a blue that made the fenman's red-brown freckled features cerulean and dyed the backs of the grazing flocks. After leaving this water-way and its magic beauteous glances, we drew up to a decaying landing-stage, startling the sporting fish over the green hair-weed, ere we took our way to the sea, which the dead ragworts and rabbit-eaten thistle-roots proved we were nearing. Indeed we were close upon it, and turning into a sandy lane, the gap in the dunes gaped intensely blue before us. Beyond we could hear the cry of the sea, and not even a broody hedge-sparrow that limped cunningly before us attracted

* I have seen the Fire-Crested Wren once in the Broadlands.

our regards—they were turned towards the wet sea-beaches.

A little way down the sandy loke two little birds flew up—tom-tits, I thought; indeed, the hen, as we afterwards found, flew from blossoming gorse to gorse with the flight of a wren. And indeed I thought the bird a wren, but suddenly a little fellow with a golden crest flew from a spray of blooming blackthorn, and I recognised the brave, cheery little goldcrest. In sooth, his very tameness and confidence showed his kind; for, though brave as a lion, trusting and loving is the little goldcrest. And the seas and marshes, the gleaming ridges of sandhills and marram-crests were forgotten; the sweet little cheepings of the little pair of goldcrests were more attractive. There he sat on a gorse spray—they love the gorse down there—cheeping and picking insects from the prickly leaves. He, a mite of seventy grains in weight, a mere living atom of feathers—he, with his loving little wife, had come across the turbulent seas. What hope! what bravery! Yes, indeed, he with a young wife goes down into the deep, flying just above the waters, hopeful that no storm will arise and tire his slender wings, blowing him down into the trough of the sea to perish miserably, a common fate with his tribe, as their little bodies prove to kind-hearted fishermen who see them floating on the waves after a gale, and mutter to themselves sadly, "Poor little warmin!" for they cannot alight on the water and ride out the storm as do the stormy petrels. Poor little herring-spink! for he has to cross in the autumn with that vast crush of birds—the migrants contemptuously called by the fenmen "wheat-pickers" (*e.g.*, the tree-sparrows, rooks, larks, wood-pigeons), that the boats meet during the herring-fishing in misty autumn, when the snow falls and a north-west wind blows. They come and go by their appointed paths—larks steering east to west, flying low, just above the water. And the strong men who go down in the fishing-boats will tell you of the flocks upon flocks they meet

during the last four months of the year over the Dogger
—myriads of larks, redwings, wood-pigeons, rooks, sparrows, starlings, swallows and swifts, house and sand
martins, bramblings, homely robins, jackdaws, and Kentishmen going from and making for Old England; and they
will tell you, too, of kestrels caught and fed on herring-
gut, of wearied carrier-pigeons taken from the mast-heads
and carried into port, and of the scapings of snipe flying
high above the drifting luggers; but rarely do they catch
sight of the woodcock—he is too shy.

Then if the fleet be becalmed in an autumn fog, that makes
the vessels' lights loom ghostly and large through the damp
mists floating dreamily over the oily water, the birds get
bewildered and the fleet becomes a vast roosting-place,
and the young hands go bird-catching, capturing the air-
loving pigeons and hawks; for the pigeon roosts with the
hawk, the silent, sleepy mavises with the larks, and even
kittiwakes side by side with the little herring-spinks, that
often feed like wrens in and out of the sheaves of blocks,
or amongst the nets, picking off the sea-lice, a food they
delight in. And the captives are taken below into the
cabin, that reeks of tobacco, bilge, and tar, and forthwith
the birds become sea-sick, throwing up their food amid the
hoarse laughter of the crew; nay, even the little herring-
spink turns sea-sick. But even birds get their sea-crops,
and recover and feed, if kept below, keeping their food down
with the best sailor of all the crew.

And the next morning perhaps breaks fine and bright,
a chill autumn morning with a blue sky, and the birds leave
the ship and go noisily on their compassless way across the
watery wastes—sailing with the wind abeam—flying a yard
or so above the water.

And perchance, if your ship be near the English shore at
this season, you may see the emigrants coming towards the
land in the early dawn, and then you know for all time the
real joy of the traveller as he sights the port; for all these

vast flocks will break forth into cries of delight, encouraging each other, and mounting higher into the air, so that they may get sight of the promised land, which they hail with a pæan of exultation.

Still, though the shore and safety be within their grasp, some of the weak and feeble of mind or body will, weary unto death, sink down and alight on the sea—the treacherous sea, that swallows them up within sight of the dunes. Mavises and larks are prone to give up on the eve of victory ; but even some of these laggards, at the wet touch of the ocean, struggle up on tired wings and struggle to the shore, swimming down the blue air, dazed and drunk with fatigue, to the dry warm sands—safe, and forgetful of those who have perished. And amongst them will be found our brave little herring-spink—the coy, tame little bird—that emblem of bravery. The weak in heart should take the goldcrest as his guiding star, and the augury of birds might be of living worth to him.

CHAPTER XIV

THE WILLOW-WREN

SOME time in the second week in April, when the willows in the carr are breaking into new life, if you chance to pass that way in the soft diffused light when the tall and slim saplings shine softly brown or green, you may hear the soft sweet song of the willow-wren ; and if you look up amongst the slim leaf-blades, you may see the little greenish bird, just arrived from Africa, running about the mossy bark feeding on the flies and "little millers" (a small white moth) he came all that long journey to find. He is not shy, and if you do not make too much noise breaking through the close coverts, he will go on unconcernedly feeding, working all over the tree, hopping from branch to branch at times, resembling a reed-warbler in form as he flits to a dark corner amongst the green boughs, or again feeding like the wren or saucy "pick-cheese." You may see him better, perhaps, as he flits to and fro seeking "green-fly" under the leaves of sycamore trees, or darting at the caterpillars that hang by silken threads from the tree-branches, especially when he has young in his oven—for his nest is just like an oven ; hence he is called the "oven-bird" in the fenlands.

All through April you may hear his sweet song, and in May he begins to build on the ground in a planting at the foot of a willow, or in a clump of nettles beneath a bramble-islet, or by some grassy dike-side where he scoops a hole and roofs it over, or beneath the ivy in a hedgerow. I have seen the nest in all these places. The oven is built of moss, grass, mud, hair, and a good warm lining of feathers, upon which the six or seven whitish eggs with pinkish spots are laid. The hen enters the oven by a good-sized hole, a door nearly twice as large as the wren ; and in this warm oven the old bird sits very closely, only leaving the nest if you get very near to her, when she darts off and disappears ; but even then the nest is difficult to find, and though the birds be not uncommon, their nests often escape the egg-robber. When she darts from her nest, if you lie down and conceal yourself, and then wait patiently for some quarter of an hour, you may see her come dancing back, looking round anxiously as she flits from branchlet to branchlet, gradually drawing up with great caution, and dropping into her nest ; and then you too know where it is ; but don't disturb it, for she only lays once, and if robbed will not lay again.

But should the eggs be hatched and the oven filled with young, if you pass within earshot they will betray themselves, for they are for their size the noisiest youngsters I know, continually " weeping " for food : nor are the old birds often far away from their brood. Many a time when walking along a dike I have suddenly heard the young willow-wren calling me straight to the nest, the old birds flying up the rond a little way, looking on with beating hearts.

A friendly, tame, affectionate little bird is the simple little willow-wren. One of the joyous, sweet-voiced harbingers of spring, he is always welcome, the child-like little willow-wren. Coming in April and staying till November, he is like some of the pale flowers of the fenland—as charming as evanescent.

CHAPTER XV

THE REED-WARBLER

THE "Reed-bird" or " Reed-chucker" (for the reed-warbler is known by both names in Norfolk) follows the sedge-warbler and grasshopper-warbler across the sea ; he is the last of the three to appear, waiting patiently till the young reeds shall have grown three or four feet in height, and there is plenty of cover to hide himself and his bride; for the reed-warbler is a shy bird, and it is not easy to get a shot at him once he takes to the reed-beds, though you may see him when he first arrives, for he generally roosts in the sallow plantings for a few days on his return across the sounding sea, which is oftener at the beginning of May than not—in May, when the gold and black chate flowers deck the dikes.

From the cool shades of the sallow plantings they fly to the reed-beds, seeking reed-stems that grow in boggy places and marsh swamps, where they grow long and stand rotten, for there he finds his food, the reed-maggots, that leave their reed-cases in spring—maggots looking for all the world like meat-maggots. These are his dainties, but he does not despise insects floating on the water; and if you sit quietly amongst the creaking reed-beds, you may see him running up and down the canes hanging over the water like a mouse, feeding upon his choice insects and maggots.

The hen-birds often arrive with the cocks, and once they have taken to the reed-beds you may see them chasing each

other from reed-clump to reed-clump, or you may see the
cocks flying up in the air like butterflies, sailing down into
the reeds with outspread wings. By night, too, the courtship goes on, so eager are they to pair, and though they
soon mate, they are in no haste to squander their honeymoon—they wisely enjoy their young lives, singing sweetly
at intervals by day and night, but most sweetly at break of
day and closing-in time. Only one hour seems sacred to
them, the hour before midnight, when the reed-beds grow
silent, though some restless youngsters may break even that
solemn stillness.

If, in the bright mornings of May or long June evenings,
you push quietly into the reed-beds, and follow some watery
path bright with white water-ranunculus, arrow-head leaves,
and white and yellow water-lilies, and bordered with beds of
lush gladen brown with spindles, nodding bolders with their
crowned and feathery heads erect as plumed lancers, you will
hear, as you move silently through the soft ghost-like reflections of the water-plants, the sweet song of these birds—a
song sweeter in note and more finished than that of the sedge-
warbler, but hardly so spontaneous and joyous.   There,
amid the soft music of the water-plants, as they whisper
together like lovers at dusk, come the reed-warblers in
legion and drown the music of the breezes.   And as their
chorus dies away, a pike splashes, a coot calls, the breeze
freshens, rustling the water-plants, and the reed-leaves
crackle till you feel their hardness and sharpness; the
breezes gain strength, their passion rises, and they sigh
through the watery jungle, the reed-plants clashing against
each other and creaking as they bend before the freshening
breeze: then the reed-warblers are silent, for they love still
quiet weather; they brook no rival music.

But as the sun warms the waters the bird's passions
kindle, and the happy pair begin to weave the wonderful
nest, choosing three or four (oftener four) stout amber
reeds for support.   Beginning above high-water mark (some

2 ft. 6 in. above the tide), the birds take a length of dried wiry grass at the ends, and together they weave their cradle, filling in the outer layer with reed-feather; next comes a cushion of flaxen spindle fluff, reed-feather, and at times a few swan's feathers; then a wiry framework of dry reed feather, and last stage of all, a cosy lining of soft white cotton-grass; and, lo! the cradle is done and placed above the water safe from vermin, near food for the young and secure from storm; for though a gale blew over the reed-bed, the reed-stems stand secure as piles made of steel.*

After the few days of toil passed in building the perfect little nest the hen begins to lay her four or five greenish eggs freckled with olive. Nor is she in haste to get that over, for she often misses a morning between her layings, spending her time in listening to her mate singing his chucking yet sweet note. The song then betrays to the egger the whereabouts of the dear eggs; for your Broadsman always hunts these eggs by sound. But the egger must be alert, for once the eggs are laid the songs grow rare and more rare.

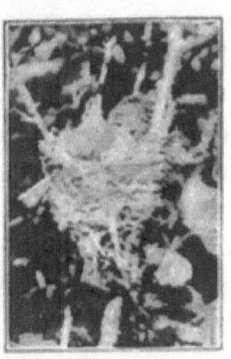

REED-WARBLER'S NEST IN A BLACK-CURRANT BUSH (Norfolk)—(taken *in situ*).

Though essentially a denizen of the level swamps, the reed-warbler at times builds in sallow plantings, and even in gardens, as the plate shows. I once found one built upon a briar and a single reed-stalk in a planting by the water, and very tame was the sitting bird—the hen on this occasion, for both birds sit. I watched the little hen for

---

* I have never seen wool, moss, or horse-hair used in the making of these nests in Norfolk, but often cotton-grass-down, which may have been mistaken for wool. Nor have I ever found a cuckoo's egg in a reed-warbler's nest,—as asserted by Mr. Saunders and others. Nor have I ever heard of marshmen finding such. In a *sedge*-warbler's nest, yes!

long before she flew up and ran away up the branch of an
alder growing from the tangled clump of briar and reed. But
the uxorious cock soon appeared, and it seemed to me ordered
her back, for she returned, as I stood, to her eggs, and sat
tamely enough, eyeing me from her compact shot-shaped nest
—a piece of confidence with which the sedge-warbler has
never honoured me, though his eggs are smaller and darker
in colour, and not so easy to be seen.

When the hot sun of July beats down upon the reed-beds,
the young nestlings are born in the romantic home that does
not rise and fall with the tide, as some aver. Then the cock
and hen are busy indeed gathering insects, running up and
down the reeds like mice, hanging in all sorts of quaint
positions as they collect maggots and insects and the embryo
dragon-flies hatched on the reeds. During this busy time
both birds keep near the young, one sitting on the nest and
uttering hoarse little notes, whilst the mate, who is collecting insects perhaps from the blazing marsh-marigolds,
answers in the same voice. Or at intervals the cock breaks
into short joyous snatches of song. At this season, too,
they frequent clumps of sallow in search of flies, of which
they are very fond, but they seldom leave the reed-beds far,
and when they choose a secluded swamp, are rarely to be
seen once their nursery duties have begun ; in short, they
keep to the reeds all summer.

The nestlings cannot fly far when they leave the cradle,
but climb about the reeds like mice in search of food, moving rapidly through the reeds, or at times nestling together
upon a reed-spray, making as pretty a little picture as one
could wish for. Should you happen to come upon the fledglings in their cradle, they will, like sedge-warblers, run out
of their nest and glide away like mice into the reeds. But
they soon grow, and when the September moon arrives the
reed-beds are again silent, for the warblers have gone across
the seas, and never a one is to be heard through the grey
cold winter.

The reed-warbler is an artist, reticent, shunning publicity, and fully occupied with its own work, as all artists should be. To him, to sing and weave those perfect nests and rear his family is the be-all and end-all of life.

REED-WARBLER'S NEST AND EGGS—(taken *in situ*).

CHAPTER XVI

THE GREAT REED-WARBLER

ONE June morning I was working through a large reed jungle, the water over my boot-tops, with a quick-eyed Broadsman, who knew every bird of the district by sight, and name too, though his nomenclature was provincial.

The sun was shining brightly, and a nice south-easterly sailing breeze blowing through the old reeds, that shook their ragged tassels this way and that, so that, as we marched through the tall crop, we grew dizzy with the ever-shifting reeds, and seemed to be walking up hill and down, though really on a flat bog. Suddenly the marsh-man stopped in his sinking footsteps, held up his finger, and looked eagerly towards an opening in the reeds on his left, and yet in front of him. I listened as I sank into the soft ooze, and heard a hoarse chuckling—a voice more like that of a large mechanical sedge-warbler than anything I could think of.

When the notes stopped, the marshman crept forward stealthily in the direction of the sound, whilst I stood still, now over my boot-tops in water, looking silently after him. As he neared the opening, I saw him looking keenly round the yellow waste of stalks, and suddenly he turned and beckoned to me eagerly.

I stole up to him, and he pointed to a broken spray of reed some thirty yards away, and whispered eagerly, "What be they? Look yonder—reed-birds as big as mavishes." I looked in the direction indicated, and saw at once the plate in Lord Lilford's book—only alive; they were undoubtedly

great reed-warblers, my glasses confirming the diagnosis.
Both birds were sitting on the reeds, as if resting, and we
stood silently watching them ; but they did not rest long, but
flew off into the reed after the manner of a reed-warbler.
And these were the first and last great reed-warblers I ever
saw alive, and they were as " big as mavishes."

A native collector told me afterwards he was sure they
had bred two miles from there the year before.

CHAPTER XVII

THE SEDGE-WARBLER

BEFORE the last rime-frosts have gone, the brave little cock sedge-warblers come across the sea in small parties, and any day after the middle of April you may awake to hear either the cuckoo or this quaint little fellow, with his hoarse canary-like voice, singing from some little clump of reed in a dike, from some islet of bramble, or from some hedge near by, his song, *chuck, chuck, chuck, chuck, chuck, chuck*, changing his note to a *click, click, click*, and then to a soft *wheet, wheet;* and if you watch him, you will see him fly up into the air, then spread his wings, and sail softly down, like a great brown moth, into the stuff. Should you whistle to him, he will try and imitate you, and, should you stone him, he will sing the more cheerily. Blow high or blow low, come rain, come storm, he will sing, by night and day, his never-ceasing little carol, so characteristic of the Broadland —that song which, when you hear it far away from Norfolk, recalls vividly the lazy rivers and idle lagoons.

About a week after the males arrive the females appear— little brown shy birds, to whom the males sing, chasing them along the reeds and hedgerows, flying up singing into the air some twenty yards, and sailing prettily down, still sing- ing and generally showing off; but should you approach, they will drop into the stuff on sight of you, and sing on in a subdued voice, for they are shy or suspicious. Should you approach too closely, a hoarse *tut-tut* greets you; and if you flush them, they go flying down the stuff or hedgerow, and dart into a snug corner some thirty yards from you, at once resuming their song; for they are saucy, provoking, jolly little creatures, and I prefer their song to that professional musician, the reed-warbler, as he has too much "side" for me.

Towards the end of May, when the last frosts have gone, the little pairs begin to build their **nests,** preferring sallow bushes in a sedge marsh; but hedgerows near water, or clumps of reed in half-choked dikes, or clumps of rush suit them, if they cannot get a sallow forest on a sedgy plain; and at this season you will hear the cock's bright song by day and night all over the flatland, for these birds are four times as numerous as the cock reed-warblers, whom they most resemble.

Should his nursery's aspect be too exposed, he will hide his nest behind a rush screen, and once or twice I have known him hang it from a branch of a bush, or two or three reeds, after the manner of the reed-warbler, but differing in this, that he always hangs his nest from one side, and never builds it about the reed stalks and branches, as does the reed-warbler. What he uses to build with depends, as in the case of most birds, on what materials are handiest; also, the nest is placed at various heights from the ground, according to circumstances. I have known it to be built a few inches from the ground, a foot, two feet, and two feet six. In any of these positions you may see their long nest, but most frequently in stuff half-laid down, with its contracted rim, built of fine dead grass externally, and lined with horse-hair and wool, and reed feather. And when the little pair have built their cradle, which generally takes them four days, the hen begins forthwith to lay her sedge-green eggs, speckled with hair-like markings, the colour of sedge-flowers —the pair lurking near the nest after the first egg is laid, the cock still flying up and crying as of yore. As a rule, every morning an egg is laid, until five lie cosily in the nest litter, and then she begins to sit, he singing joyously near by from a reed-top, hanging this way and that, and dropping down if disturbed. I don't think he sits at all, as does the reed-warbler. She is artful enough to manage it all, however; for, should you flush her from her nest in the open, she will feign to be wounded or lame, like many another, and lead you astray, as mayhap other ladies have done before.

In the fulness of time, when the young maws are born, you see the old birds busily gathering maggots, caterpillars, insects, moths, and midges, chiefly from the top of the water, ere they hurry back to the nest. You may at times stand on a marsh and see four or five pairs flying to and fro feeding their hungry young, for it is very easy to watch them to their nests; their journeyings to and fro are so frequent, that if you lose sight of the parents the first time you see them, they soon reappear. Should you approach the clump of sallow where the young are, the old birds will come to meet you, and fly around uttering a short *chuck*, hoping to lead you astray; for they are jealous of their homes, and you may often see rival pairs chasing each other away from their particular sallow islet across the green sea of rush or sedge.

Sometimes, before the young are half fledged, they will come from the nest to hunt for food, returning again; but this is rare. The young are difficult to catch as a rule, escaping along the ground and through the stuff, coursing rapidly along like mice, and seldom will you see them in the nest fully feathered, though at times you may, when they will at once flit from their cradle like mice, and scatter through the herbage.

Soon these young birds begin " to do " for themselves, and the parents forthwith begin to build a second nest and raise a second family—a piece of work only undertaken by the reed-warbler when he has been robbed of his first family.

In fine seasons I have heard the little fellow singing as late as the last week in October, but never later than that, for the cold grey mists rising from the rivers and lagoons like ghosts, warn him it is time to go where the sun is warm and insects abound.

Brave, joyous, manly little fellow, with your endearing manners and pretty ways, long may your race flourish, though oft-times your eggs are sucked **by the** pilfering mouse and relentless weasel.

CHAPTER XVIII

THE GRASSHOPPER-WARBLER

THIS shy, mysterious bird, the "razor-grinder,"* as he is often called in the Broad district, is oftener heard than seen.

When the chate is in bloom the grasshopper-warblers come over in small parties, which scatter at once over rush-marshes dotted with islets of sallow, bramble, and reed; for they are not such water lovers as are the reed and sedge warblers — nor are they such musicians. Indeed, unless you be alert, you may hear the earlier homing sedge-warbler, and the later arrivals, the reed-warblers, and not have heard the grasshopper-warbler at all. Yet he has often sung to you (for he arrives between the other two songsters), but you may have thought him some insect, so low and mysterious is the chattering coming from the marshland. By night and day, too, he sings when the days grow long. On the other hand, most people would not know he existed were it not for his song, for he is the most skulking of birds, a mysterious and reserved little minstrel, who shuns publicity.

But if you know the marshlands, and frequent a marsh of "laid-rush," where brambles and sallows grow in clumps, you may at daybreak and closing-in time hear his mysterious song; and if sharp-eyed, you will see him sitting on a bramble spray a few feet from the marsh. But should you approach even at a gallop, he will drop into the stuff like a stone, and you may search in vain, though, if he have dropped into a bramble-bush, and there be two of you with sticks, you may succeed in flushing him, when you will see that he is a bigger, darker, more broad-tailed bird

* Never "reeler."

than his brother warblers—a bird resembling a hedge-sparrow more than a warbler. But you must be quick, for he will not fly far, and if you pursue him he will dive into the stuff, and you may search the marsh all over—never will you see him again, for he is as alert and quick as a marsh-mouse.

All night through flowery May you may hear these " grindings " of his at intervals—the song sometimes lasting for half-an-hour or longer. When the cotton-grasses are ripe at the beginning of June, the pair begin their nest in the soft, green moss on the marsh bottom, hidden beneath the laid-rush, or in fairy forests of cotton-grass. There, in dusk seclusion, the nest, resembling that of the lark, and made of dried grass and roots, is placed; and so cautious is the bird, that should the nest be disturbed by the curious, she will desert it, although one marsh-mower once mowed a nest up in his swathe, the bird starting from the marsh at his feet. He replaced the swathe, never touching the nest, and tried to chase the bird, and "muddle her out;" but she seemed to sink into the ground. However, he returned to the nest some days afterwards and found two eggs, which he took, leaving her to start anew, as she does if once robbed, as does the reed-warbler. These birds have runs to their nests like marsh-mice, and if you flush them they will fly a little way, go down to the marsh like a stone, strike one of their dark arcades, and track for yards under the stuff as securely as a mouse, coming back later on, when you have in despair given up the search, to their four, five, or six spotted, pinkish eggs. But if you are determined to find a nest, cut a long stick, and watch for a bird "grinding," for they always grind near their nests; then walk steadily towards the grinding bird, beating the stuff with your stick, but looking steadily some twenty yards ahead of you, and you may be rewarded by finding one ; but your chances are small, and these nests are seldom found, save by men mowing the rushes for litter. You must be content to see them sitting on a spray of bramble or sallow some feet above the marshes, "grinding" with their heads set up straight, and tails hanging straight down ; for you will find as you approach, the shy bird drops on to a lower bramble, and finally, as you draw nearer, disappears in the stuff.

YOUNG GRASSHOPPER-WARBLERS AND NEST.

But you may oftener see the young birds, for though they cannot fly out of the stuff, an active man can fall suddenly upon them and catch them ; yet it is a difficult feat, and an impossible feat when they are fully fledged, for then they leave the stuff for the clumpy islets of sallow, bramble, and sedge. But these birds are oftener seen than caught, because they too have grown more active and knowing, and drop into the stuff like stones, and are lost to view.

After harvest is over they are seldom to be heard "grinding," though at daybreak and throughout the summer they are to be heard daily. But July is their nosiest month—July, when the marshes are gay with ragged robin, blue oxytrip, meadow-sweet, cinquefoil, red docks, and yellow rattle. Then at daybreak, as the mists are clearing, you may hear the birds "grinding" amongst their coppice islets in that marshy sea, like some huge cicadas, for three, four, ten, or even twenty minutes by your watch, stopping merely to get breath, and going on till the mists clear and the garish day exposes them to view, when they rest for a space, beginning again at ten in the morning, "grinding" through long spells with their heads thrown well back and their eyes looking all around them over the green marshland, the songs rising quicker and sounding shriller until the birds stop for a moment, the intervals being of different lengths, then continuing it for over an hour with a few momentary stops, the song recalling the winding of a spring steel-tape in different lengths, now stopping suddenly for a moment, now being pulled out quicker and quicker, then suddenly stopping. And so on at intervals the songs —those mysterious voices—go on by day and night till the end of summer, when the birds go to "grind" music upon some far distant marshes.

GRASSHOPPER-WARBLER'S NEST AND EGGS.

CHAPTER XIX

HEDGE-SPARROW

THIS quiet, tame, sober-coloured little bird goes in the Broadland by the name of the " Hatcher," perhaps because he sometimes "hatches off" the lazy cuckoo's egg—though neither he nor his nest are favourites of the cuckoo. His short robin-like song is one of the charming, hopeful voices of early spring; and when in February you see the cocks chasing the hens across the marsh-grass, your heart is delighted, for when the blue eggs by the hatcher are laid the cuckoo will soon be over.

Then day by day you see them lurking about the bare hedgerows, working their half-spread wings in jerks, following each other. A little later after this courting chase— towards the end of March—you may see the pairs feeding together by some gorsy islet that rises from the marshland —that is their honeymoon.

For soon the little nest is begun, for preference, in a low gorse bush already breaking into bud, or in an "ivory bush,"* or more rarely in old thorn faggots stowed by the millman's door. Early in April the well-known little blue eggs lie naked to the skies. And when the hen begins to sit, if you chance to pass that way and flush her from her nest, she—full of deceit, like many "simple," homely folk— will mayhap lie down on the road and spread out her tail, or sit back on her spread tail with her partly spread wings fluttering, pretending to be fatally ill or mortally wounded.

* Ivy bush.

But do you run up and attempt to catch her and you will see she will at once gather herself up and fly nimbly off to the protecting hedgerow, awaiting your departure from her district.

At this season too the cock shows he is not averse from joining other sparrows in robbery. You will find this "quiet" simple little soul pecking furiously at your young onion-beds or some other sweet seedlings, or he will be playing havoc with your budding bushes and trees; and if you could accuse him green-billed, you are sure that, like the "simple" girl, he would plead "it was such a little one," or "only the first time."

And when the equinoxes fill the North Sea with herring fleets, these birds spread over the country-side, living through the winter by the fenman's or farmer's door, or else they feed with the rooks, starlings, and larks and titlarks on the melting patches of snow in the bleak landscape.

And so this insignificant little bird lives its life, never doing anything to be "spoken on," adding no beauty of song or bright patch of colour to the landscape, but leading a "quiet life," practising the arts of deceit when necessary, stealing with a deprecatory air, and passing through life as do a host of "homely" and petty-minded people—as "respectable."

CHAPTER XX

THE REED-PHEASANT

LONG before the frogs awake from their winter sleep, long ere the peaty dikes are green with the spear-like tips of the reed-colts, the grey-headed, tawny buff cock-birds with their black moustachios begin their low *chings* through the tossing yellow reed-beds, where the gusty wind plays in and out, casting ever-shifting shades and lights that dazzle and stupefy the fenman on egging bent; for the reed-pheasant's eggs and skins are beloved of the collector, that fatuous gatherer of unconsidered trifles. If the weather be open upon St. Valentine's day the birds begin to pair, and you may, as you glide along the silver water-ways, see a little band fly up into the ambient air as high as a wherry's vane, *chinging* as they fly, then suddenly stop, turn, and dart down again into the yellow reed-bed, as if shunning the light. Before this augury the taciturn fenman by your side will say slowly, " They'll soon be laying now." And he is right; for immediately follows the playtime or honeymoon, when the happy pair fly about the " thyte (thick) reed," plucking reed-feathers and tossing them wantonly hither and thither.

At such seasons, if you lie silently by a green rond ablaze with kingcups bordering a reed-bed, you will see the handsome cock-bird, grotesque with moustaches, flit to and fro, and run up and down the reeds like a mouse, for in pairing these birds never seem to quarrel or chase each other in that aimless manner peculiar to most birds. Indeed, I have known two birds to lay in one nest. Then, too, as you wander through some of the reed-jungles, you will find in the chate several cocks' nests, as the countrymen call them, for the cock reed-pheasant is very fond of making these abortive attempts at nest-making.

A week or so later, if you go into the swampy jungle (at the end of March), and look carefully through those parts where the quaking bog is knee-deep with chate, soft rushes, or beaten-down gladen-stalks—for they invariably choose a dense undergrowth of one or all of these marsh crops wherein to lodge their nests—you may be on the look-out for these curious birds running up and down the creaking reeds, or making pheasant-like flights through the amber stalks, *chinging* as they go. Once you hear the cock's metallic cry, if you be nesting and are experienced, you will suddenly stand perfectly still, and strain your eyes to catch sight of the mice-like bird running up and down the reeds. As you stand silently in the soft ooze, sinking in the water above your boot-tops, you may observe him run up a reed-stalk like a mouse and pick a piece of reed-tassel, and then fly straight off to his nest, uttering a metallic *ching, ching, ching*. Then start off through the reed-jungle as fast as your waders will permit, keeping an eager eye upon the reed-bush where he alights. If you are fortunate to have come close to the nest, and possess your soul in silence, you will soon see the little birds working about the reed round about you, running to and fro like mice. In such plight you may know you are too close to the nest; so move quietly off deeper into the jungle, and crouch in the dry, crackling under-growth of chate, and watch stealthily as a tiger watches its prey. Presently you will see the cock-bird betray the nest, for he will begin plucking dry reed-feathers and dropping them over his cradle. If experienced, your heart will jump into your mouth as you bound forward and search amongst the dry chate about from one to two feet from the hover bottom, the old birds meanwhile having run off in the stuff like little mice. There, cradled in the undergrowth of sedge, you will, if fortunate, find the long nest, about nine inches in depth, deftly woven of the dried bottom leaves of the reed, and lined with reed-feather; occasionally a coot's or swan's feather being thrust jauntily into the reedy boat-shaped cradle *pour s'amuser*. As you part the dry and crackling chate, you may catch sight, for a moment only, of five yellow bills peeping from brown heads, lost in a mass of blackish feathers tipped with buff, and you may hear a *cheep, cheep*, and lo! they have bolted out of the nest and run like mice away into the stuff. Still you may know them to be near, for the old birds will come working and calling round you; you can hear them *chinging* and making a peculiar grunting kind of noise. That is a signal for you to move off into the deep reed, for they will surely return to their nest even when they can fly—indeed, young reed-pheasants are the only birds who do this—and they will return four or five times a day. Having taken your mark, you draw off, and return upon a rainy day or early in the morning after a heavy dew, and you will see the old birds hunting under the sheaths of the reed-stalks and at the nodes for the maggots where-with to feed their nestlings—for they love damp reed to work upon. Spiders, midges, and the seed of reed, too, are found in their bill of fare. Should you, however, be of a predatory nature, and attempt to take the nest rashly from its moorings, it will fall to pieces unless you be cautious to tie it together beforehand; but 'tis safer to cut the hover from the reed-bed—nest, chate, and all—but best of all it is to leave the long bulky nest where it is and go search for others.

You may find as many as nine eggs in a nest, but five is the more usual number; and they will, if robbed, build five or six nests in a season, moreover, and not far from the spot where the first cradle was lodged. Nine days suffice for those little architects to complete a new home, which is nearly always built of reed-leaf and feather, but I have seen nests built of litter and lined with fine grass. At times they will not wait to build a new nest, if robbed, but will drop their eggs where the old nest rested, and build a new nest over them. I have known fenmen who have frequently found old nests with eggs woven into their bottoms—cellared, as it were, and such happens with the great-crested grebe. The fenmen who gather the nestlings mow a circle round the nest before the eggs are hatched off, and net the place with an old piece of herring lint; for the birds seldom build over water, though they dearly love a hover that rises and falls with the tide, and perhaps that may account for their long nests. And when the hatchlings are a week old the bird-catchers drop into the reed-jungle and make a dash for the nest. The young birds tumble out like mice and make for the ground, and the fenmen catch them in the mowed space before they have time to reach the protecting reed-brakes, and afterwards they are reared by hand and kept as cage-birds, and they are wonderfully fast-growing birds. Indeed, the eggs are hatched in eight days and the young can fly in a fortnight. Should you not have patience to loaf about the reedbeds, however, and wish to discover a nest in a lightly cropped marish swamp, take a long light pole in your hand and start into the swamp, being all eyes (looking ahead) and ears, listening for the *ching, ching.* March along, looking well ahead and beating the stuff, and you will be sure to put them up, if there be any birds in your sparse jungle. And then be careful you do not crush the eggs or nestlings with your heavily-booted feet. And be sure you choose a *still,* bright day for your work, or never a reed-pheasant will you see. And you may find their nests as late as July. And be sure, also, that the cock-bird, who often begins to sit on two eggs, will be sure to betray the nest if you let him. Especially is he over-anxious and fussy about the first eggs; indeed, his behaviour must be a thorn in the side of his plain mate. In winter the reed-pheasants gather together in flocks, each numbering fifteen or twenty, and you may see them rise from the reed on a bright winter's day, *chinging,* flying up some yards into the blue, and suddenly throwing themselves down headlong into the yellow reed-bed to feast upon the insects therein. They are cheery companions to the solitary reed-cutter, as he works boot-deep in the icy water. He often sees them run along the fallen amber stalks, moving like a wagtail, with tail held straight out behind, picking insects from the water. And the fenman knows they will build close by, for they never wander far from the place where they were bred and born.

The reed-pheasants are hardy, sociable little fellows; indeed, in nesting time they often, like red-shanks and pee-wits, build close to one another—one "coys" the other, as the solitary fenmen say; but the hard winter of 1890–91 was very fatal to these little fen-birds, and the following spring but two nests were found in a district most dear to them. Yet they are not extinct, and may still be seen in the Broads wherever the swampy crops grow thick in winter and sparsely in spring. And, like the fenman, they are mysterious, ever seeking the seclusion of the reed-bed. The reed-pheasant is, after all, the last link with an earlier age, a period when the fenman lived in inaccessible morasses, with no other companions than the flickering will-o'-the-wisps and the watery tribes of birds whose strange voices filled his soul with a native poetry that increased his natural melancholy and superstition.

REED-PHEASANTS AND NEST.

CHAPTER XXI

THE LONG-TAILED TITMOUSE

Or "titimouse," as the fenmen call this moth-like bird, always associates itself in my mind with the humming-bird. Any one who has seen a living gem of a humming-bird hovering moth-like about the trumpet-shaped flowers of a plantain or banana-tree in the tropics, cannot fail to compare it to the hovering of a long-tailed titmouse about an ivy-covered tree in summer-time when in search of insects. Moreover, the face of the bird resembles that of a large moth in its roundness, beady eyes, texture, and general expression. Altogether he is an artistic creature in form and colouring as well as in habits, for he builds one of the deftest and most beautiful of all bird's nests. The silvery lichen-covered home, woven amid a large branch of ivy, or in some scaly-barked gorse bush, is a little masterpiece, although perhaps rather inconvenient for the mother, who has to sit with her long tail folded over her back, her feathers resembling a Spanish mantilla over her little head as you peer into the nest.

Though by no means common in the Broadland, the long-tailed tit is not rare, and often builds there, generally returning to the same neighbourhood to nest year after year. I know of one garden abutting on a Broad where a pair have nested in the ivy round an old elm-tree for several years, laying their dozen or sixteen eggs regularly. And there on a fine warm day in July you may see the young birds flying about the trees looking for insects, and calling to each other with their childish *wee, wee, wee ;* and later, when the reed is ripe, they frequent in flocks the tall trees by the river,

with the " pick-cheeses " and redpolls, flying about following
one another round the tall bare tops from tree to tree, like
a flock of giant moths, and even at a long distance you
should know them by that paddling flight of theirs, for they
seem to lie on the air with their tail and body, and paddle
with their wings. You may distinguish them as well by
their long tails. But, like the giant moth, 'tis a bird not too
commonly seen, nor are its simple habits too freely divulged ;
for during the nesting season they are very shy and skulking, and one may be surprised by suddenly finding a young
family flocking about a familiar corner of the garden where
no nest was suspected, or even old birds seen. And yet
you might think they would be more numerous, for sixteen
is a common number of eggs ; but I have seldom seen more
than five or six young birds together when the cradle is
left ; perhaps they perish from overcrowding in that bottle-
like nest. And the fenmen speak of two kinds, the blue
and pink long-tailed titimouse, describing the blue as the
bigger bird.

CHAPTER XXII

TWO MORE TITMICE—THE GREAT, TITMOUSE AND THE COAL-TITMOUSE

I HAVE rarely seen either of these birds in the Broadlands, and from inquiries I find they are by no means common. The great titmouse is locally called the "Bee-bird," from its habit of eating bees in spring and autumn, when they are somnolent and there is little other insect food. I know of one of these hatching off six young in a hollow tree, the nest being a yard below the hole in the tree, and therefore not to be reached by the grasping urchin. These six nestlings were fed upon flies and moths that were captured in the ivy growing round an old elm. When the old hen was sitting, if you approached the nest she hissed "like a wiper," as the old fenman in whose cottage she built expressed it. An old mole-catcher told me they are some-times called "saw-sharpeners" in the building season, from the well-known and peculiar grating noise made by the cock.

The "coal-titmouse," as the coal-tit is locally called, is seen about the reeds now and then, or more rarely amongst the alders; but I have never found his nest in this district, though it cannot be unknown, for the birds, though not common, are by no means rare, although all one sees of these two pied tits is a flash across the land-scape or a glance from a reed-bed—a mere glimpse here and there, of which the memory alone is left.

CHAPTER XXIII

THE BLUE TITMOUSE

AROUND some of the Broadsmen's homes grow tree-mallows, or "pick-cheese trees," as they are locally called—the seeds of this plant, called "pick-cheeses," bearing a faint resemblance to a cheese. The blue tit is extremely fond of these pick-cheeses, whence he has been called locally the "pick-cheese," and thus is explained an etymological problem that has long puzzled the ornithologist and philologist.

And a brave little bandit is this blue-headed bird. Indeed he is as beautiful as he is brave, and as charming as he is beautiful. He is well armed, too, as you may test by letting him get a peck at your finger; but look out lest he carry off a bit of flesh. And in the early spring, when the fruit trees are breaking and the sleepy bees are recovering, the blue-headed tit is busy playing havoc, eating the somnolent bees and pulling the hearts from the buds. A mischievous bird is he in a garden, nearly as harmful as the thieving sparrow, where the year round you may see them "in twoses," as the Broadsmen say—and I often wonder if they pair for life.

And when the slipshod nest of twigs, grass, horse-hair, wool, and feathers is being built in some hole in the wall or tree, or in some crack in an old barn or mill, occupied year after year, you may, if you like, test their perseverance. Do but pull out their new nest, and they will build again; or, if you catch them building in a hollow tree, go day after day and cut the door of their stronghold larger and larger; still will they persevere in their labours, and go ahead with their cradle; and should you reward their perseverance by letting them alone, the hen will lay her eight little eggs and forthwith begin sitting.

And if you bide your time till the eggs be hatched, and then go to the nest, you may have proof of their bravery, for they will boldly attack you, fighting with you with bill and claw: in sooth, you may catch the cock thus, so eager is he for his young. Indeed, so furiously will they rage, that if you have a spark of sympathy in your character, you will give the little blue-headed captain an honourable truce, and leave him until his young be reared, and eight more little blue-capped warriors are sent forth to fight and steal your currants, or in cold winter feed upon the beef rib-bones you hang forth in your garden to sustain him and his kin; for if he does much havoc he does much good, for his young family is always fed upon cankers culled from your garden plants, or moths, which he catches after the manner of the fly-catcher, and in grey winter he searches over your rotten bark and delves out insects upon which he feeds, hanging now back-down, now sideways, now standing haughtily on a branch, as he threatens a too curious young sparrow who would fain share his food with him. The "pick-cheese" never leaves his birthplace far; indeed I have known them to sleep in winter in the very place wherein they were hatched —a hole in an old mill.

And when the leaves have fallen from the trees and the marshlands are bare, you may perchance find them feeding upon a dead horse, whose carcase has been dragged forth from some dyke; and as you pass along the soppy land you may see a flock of these blue-headed rovers on the dead horse, like blue-bottles upon a corpse, though they eat maggots whilst the flies blow them, and therein lies one distinction.

Or later, in the dead of winter, when the flocks and herds have been driven from the marshes, you will find the blue "pick-cheeses" in the alders by the grey river searching for insects and seeds, or in a warm osier carr upon some bright winter morning, when the sun has melted the rime on the reeds, you may see the little "pick-cheeses" running about the dripping reed-stalks in search of insects and seed; for, in sooth, they are hard pressed at this season of the year. But they are hardy and brave, and that spells survival, as his little song informs you in early spring as you look up through the quaint decorative alder catkins and see the little blue-cap far outshining the azure itself in brilliancy—a blue star in the blue—a worthy crest for so brave and beautiful a little bird. Long may he live, bandit though he be!

CHAPTER XXIV

THE WREN

THE tomtit, as the Broadsmen call this pert, childlike little bird, always brings an affectionate smile to your face as you see his hopping, plump little body flitting over the bank, or running along the branches of a leafless tree, stopping every now and then to sing his loud-voiced song; for, though his is a little body, he has a mighty and pleasant song.

And very early in the year, before the winter snows have melted, you may see him courting, singing loudly from some tree or bush, flying up in joy some ten yards into the pure air, singing as he flies, then descending prettily to his perch like a plump moth or butterfly.

And no sooner has he taken unto himself his little wife (generally early in April) than he begins to construct those mysterious cock's-nests, in some leafless, gnarled thorn-bush, some old wall, heap of faggots, reed-thatched shed, bank, or ivy-covered elm—for all these places are dear to him for nesting, though at times stranger sites are chosen. The most beautiful nest I ever saw was placed amongst some yellow reed-stalks growing in a pulk in the heart of a coppice abutting on one of the Broads. And a lovely little picture the green mossy nest made, as it lodged without fastenings amongst the amber reed-stalks, so softly lighted beneath the budding greenery of the planting.

And when the mossy nest, horse-hair lined, is finished, the little hen takes a respite from her labours; for she is in no hurry to lay her eggs, often waiting several days before the first little egg is deposited in the chosen nest, for at times they build two or three before they are satisfied. Yet they do not always make a new nest, but occasionally lay in an old cradle.

And when the ten little eggs are laid, the little hen begins to sit, and if you disturb her she steals off like a mouse, and it is reported will not return if so much as your finger be thrust into the nest. But this statement, like many others of the same kind, is misleading, for I have often tried the experiment. Some will desert, others will not, even after some eggs have been taken; and upon one occasion I took out every egg and replaced it again, and still the hen returned to her nest, and hatched off the little fledglings, which she and her little lord fed upon insects, worms, and caterpillars, especially those found upon the fenmen's gooseberry bushes; for a pair of tomtits with a family are the best destroyers of gooseberry cankers I know, and should therefore be encouraged in every garden. In white-hot July, too, when the second brood is being reared, he is ever busy in the garden grubbing amongst the insects for his young. And when the young are growing, should you draw near the nest, they will come tumbling out one after the other like children bounding out of school, and the anxious parents hop about on the bushes calling. And when they have left the nest, you may see them, aye, to the number of ten, sitting in a row on a bramble spray sunning themselves like playful little children, and a pretty little group they make.

And when the autumn rains and frosts and gales have stripped the trees, you may see your little friends—many of them having braved the sea-voyage—at eventide skulking behind old stumps, or round about pollarded trees, where they sleep—sometimes as many as twenty in a hole, keeping each other warm. A hole in a reed-thatch, or even an old sparrow's nest, is used at times for the same purpose.

And then the boys go forth to stone the "king of birds," as they call the wren in Norfolk—one party going on one side of the leafless hedge, and another ruthless gang following on the other. And the tomtit fares badly, for he never leaves the cover of the hedge, like other birds, but goes flitting along, hiding behind stumps, running into rat or rabbit holes, or hiding in a hollow thorn-ball, occasionally escaping the stones. At times a dog is brought along with his tormentors, for since Tommy has scent, a dog will snuff and follow, as eager for the hunt as the boys themselves. A few years ago the boys could sell Tommy's tail feathers to a Norwich chemist—for tying flies, I presume.

But the wiser tomtits hang about the stacks and remain near the farmhouse, where food is more plentiful when the stack tops are white and the pump is frozen.

And so he lives his simple life, a joy to all, a dear little childish co-mate in our journey through life.

CHAPTER XXV

THE TREE-CREEPER

THE fenman often sees a little bird running up his apple-trees or the old stumps in the hedgerows, climbing up a few feet, then stopping to feed hurriedly, and then hastening on again, supporting itself on its tail and with its sharp claws. It is the tree-creeper.

There is not much to know about him, except you may see him feeding on the tree-trunks, running up like a mouse either when trees loom large and green in summer or look small and bare in winter. Sometimes he travels round the tree in a spiral, especially if the weather be fine ; but when the wind blows you will see him work under the lee of the trunk, for most little birds dislike wind—it ruffles their plumage ; indeed, many water-fowl sit head to wind chiefly on that account.

Though the tree-creeper builds every year, to find his nest is a rare matter, for 'tis generally placed in some dark little cavern whose mouth is concealed by ivy or other greenery. The only nest I have seen came from a hollow apple-tree, and is a curious structure, composed almost entirely of moss and old cobwebs, those dust-powdered old draperies that hang from deserted barn rafters, the whole lined with grass, upon which rest the red-spotted eggs, never to be mistaken if once seen.

A mysterious shy bird, rarely giving one a chance to study his simple life, the tree-creeper is the bark-bird : as constant a companion to the old tree as the moss and polished ivy.

CHAPTER XXVI

THE WAGTAILS

A DECORATIVE, elegant little family are the wagtails—
birds that should be dedicated to the Muse of Painting.
Whether on a winter's morn you see the common wagtail
darting to and fro—tracing lovely patterns across an old
wall as he hawks for drowsy flies dozing in the sun ; or
whether it be the pied wagtail tripping daintily along a
silver dike, snapping dead flies floating by the shore ; or
whether it be a graceful yellow wagtail, flying elegantly
above a water-ranunculus petalled dike in early spring, his
yellow and black and white tail cutting the air flamewise ;
or, lastly, whether it be a lovely blue-headed wagtail running
daintily along the green top of a marsh wall, stopping to
dart swiftly at the flies collected in a horse's track—all are the
same—all are elegant, graceful, dainty, lovely to look upon ;
and, as the nightingale is the voice of the English landscape,
so is the yellow wagtail the bright and graceful jewel of our
low-toned English landscape.

The yellow wagtail, or " wangtail," as the fenmen call him,
arrives in Norfolk early in April. As you are walking over
the grassy marshes, bright with the lazy dike-weeds, now
in flower and gay with Parnassian-grass petals, you start
as a flash of yellow light gleams athwart the lush green,
and you smile, and your heart is glad, for 'tis your first
" yellow-hammer," as the Broadsmen often wrongly call him.
Already the many pied wagtails have arrived, but there is
still the blue-headed one—that rare visitor to the Broads—to
follow. In a few days, the marshlands and dikes will be

alive with the beautiful yellow birds, then a few days later
follow the hen birds, of soberer hue, possessing a greener
plumage and a more staid and less graceful demeanour than
their mates, as you may see later.   They walk quickly along
the walls, darting at the flies rising from the steaming dung
left by the horses and cattle.   With them may be seen at
times starlings and even herons.   The yellow wagtail's soft
ching-i-u, ching-i-u—a call indistinguishable from the voice of
the pied wagtail—sounds sweetly amid the cries of the lambs.
Their courtship is brief and mysterious.   You see pairs sitting
on the clumps of leafless bramble, on docks, and at times
on thistles, dear to the goldfinch, and, though a shyer bird
than the pied wagtail, they are by no means timid.

In June, when the grass-marshes are yellow with crow's-
foot and red with ragged-robin, and the holls are decorated
with elegant masses of flowering hawthorn, the yellow wagtail begins to weave his grassy cradle on a mossy marsh
bottom near the water-side where the milk-white water-lilies
are just beginning to unfold their shapely cups amid the
smaller florets of Parnassian grass and the large-leaved
burdocks.   The beautiful little couple will often choose, too,
a moist marsh, fragrant with mint, and pied with the tossing
plumes of cotton-grass.

If you watch them building, you will see they have chosen
a "hill"* where there is a scant crop of pin-rush and chate.
There on the drier side they work a hollow with their claws
and breasts, a little cup-shaped snuggery, larger than a titlark's cup, which they line with dry grass-stalks, but they do
not weave their cradles so closely as the titlark, nor yet are
their nests so symmetrical.   The titlark is the more metho-
dical architect, but the wagtail is the more finished artist.
Next to the grass he sometimes places a warm layer of
non-conducting wool, torn from his co-mates of the pasture,
whilst inside of this wool he weaves a lining of horse-hair
picked from the marsh-herds, crowning the work with a

* A "hill" in Norfolk is a dry patch of very slightly elevated marsh.

spray of swan's-down or sable wool. But sometimes he
must have a roof to his cradle, so he pulls over a few stalks
of dead rush, bending them into a beautiful Gothic arch,
lacing their springy ends into the spongy peat with. dried
reeds bent prone, all sewn as lightly as wickerwork to the
spongy peat with dried pin-rush threads.

And the door of this little house faces the east, in order
that the sitting bird (for they take their turns at incubation)
may behold the gleaming sunrise of a summer's morning
mayhap, but more likely that the great heat of the day may
not play upon its six tender nestlings when in the fulness of
time they are cradled—the yellowish-brown chicks that have
begun their battle for life. And you may sometimes know
the exact position of the nest, for the birds often hover over
the nest before dropping into it, and you may hear the
peculiar monotonous note of the sitting bird—a note that
never varies. Often, too, you may see them, on leaving the
nest, stop and arrange their dainty feathers before ranging
for food. And when the blue waters are white with water-
lilies and red with pond-weed, the young birds leave the
nest at the end of June, and you may see them on dewy
mornings searching over the marshes for flies and moths,
which they love dearly. All the summer season they follow,
in large parties, the stalwart marsh-mowers from daylight
to sunset, as they sweep down the coarse marshy crops,
disturbing myriads of flies and moths, upon which the yellow
and green families feed, regardless of the labourers.

Amongst these followers of the fenmen, too, you may occasionally see a slenderer, shyer bird, with a beautiful blue head
—the blue "wangtail," as the mowers call him; but he is rare
—to see six in a season is to be lucky; and to find a nest
very rare indeed. But such was my luck upon one happy
May—a nest holding eggs yellower and larger than those
of his yellow brother. But rarely indeed is he to be seen,
though my wherryman was stupid enough to shoot the
father of the family ere the eggs were all laid.

All through the long days of July and August you may
see the yellow wagtail taking long flights, at times crossing
the green reed-beds, as they range from marsh to marsh,
and in a fine autumn they linger with us through September and October, and I have seen a few as late even as
November; but that is extremely rare, and the autumn must
be fine indeed—an autumn such as delights the herring-
fisher's wife.

More often the first frost falls in October, and they are
gone across the salt seas to lands where the sun rises as
yellow as their own beautiful breasts, lands teeming with
the insect tribe of flies and moths.

The pied or "black wangtail," as the fenmen call him, is
resident, but the majority arrive long before the frosts have
left; indeed I have seen them flying amongst the peewits in
February, ere the kingcups have opened their golden cups.

A trusting bird is the black wagtail, the miller's friend,
for he dearly loves to nest in a pollard overhanging a sleepy
dike or beneath the prickly eaves of a gorse-thatched shed
where the miller fattens his winter calves. Perhaps this
shrewd and elegant little harlequin has an eye to the flies
and moths that frequent the miller's sheds in the first warm
days of spring.

A sociable little fellow, too, is the pied wagtail, for if he cannot find company by the millside, he will away to the marshes
where the fenmen are cleaning the dikes or mowing the
sere marsh crops, where he alights, throwing his tail up into
the air, bowing coquettishly. After this ceremony, you may
see him intent upon his prey and forgetful of the labourer's
presence, merely crying *ching-i-u*, *ching-i-u*, if disturbed by
meak or crome that drags forth the lamb's-tail.

Very fond too is he of alighting on a marsh gate or
decaying post, whence his regards extend over the marshlands, brown from the winter's frosts, away to the mill-shed
or pollard where he and his sober wife intend building their
large nest of grass and horsehair, that you may see him

steal in early spring from the horses' manes; for this impudent little fellow thinks no more of riding the marsh on cattle and sheep than the starling; indeed they often ride together and are great friends.

When the nest is made and the four eggs laid (less blue and smaller in size than those of his near relative the white wagtail*), you may see him hawking over the marshes and catching flies round the cattle's feet or getting worms from the dikes; and when he has gathered his load he flies away in a bee-line for his nest in the pollard, now green with leaves, for it is May ere he builds. And these twain take turns at the dull work of incubation, both sitting pretty close, but never allowing you to catch them with your hand. Year after year they will return to the same spot to rear their young; year after year you may recognise the sweet *ching-i-u* about the old mill-wheel until the cold of winter drives most of them across the grey seas to another old mill by an African waterside.

In winter you may see those that remain feeding in sober plumage along the water, clad in dull winter coats. Indeed, when the meres are ice-bound and the marshes covered with snow, you may find them by many a runlet eating what they can, and pleased with what they get.

His brother, the white wagtail, is rare—rare as the blue-headed bird; but I have on occasions seen this elegant bird by water near stone walls, and once I found his nest in a pollarded willow overhanging a sleepy dike, where his wife had laid five eggs, bluer and larger in size than those of the pied wagtail. I noticed both birds sat on them; but a greedy urchin found the nest, and seized the eggs, and my studies were cut short.

* A reviewer in *Nature* seemed to question my statement in *On English Lagoons* that the white wagtail and blue-headed wagtail had been seen in Norfolk. Mr. Fielding Harmer, in *Wild Life on a Tidal Water*, records the white wagtail's appearance in Norfolk, and the specimen of the blue-headed wagtail my wherryman shot was sent to Mr. George Grimsell of Reedham to stuff.

And would that some English Hokusai would paint a characteristic panel of wagtails in their beautiful coursing flight, with fan-spread tails, across a lush marsh, yellow with kingcups and rosy with the ripe crop of the water-grass, and send it to me to keep and treasure as a joy for ever.

THE HAUNT OF THE WAGTAIL.

CHAPTER XXVI

THE TITLARK

ISLETS of blossoming hawthorn, clumps of pale green sallow, and trailing brambles rise from the green sea of the far-stretching marshland, while a few scattered windmills guard the grassy plains like beacons—marks to the lazy wherrymen gliding through the land behind their black sails, that disappear round groups of marsh-farm buildings recalling arks anchored in a green sea. Above is stretched the sky, with the soft fleecy clouds of May, and the warm moist air is quivering with the plaintive wild cries of plover and the joyous carolling of the larks; when suddenly there is a brief song as a little brown bird rises from the green sea and flits upward as far as the mill-neck, calling *tu-wheet*, *tu-wheet*, in *crescendo*. When it reaches this height it pauses, turns on its side, and drops with outspread wings, calling first *chuck-a-chuck-a-chuck*, like a sedge-warbler, and alighting, it skulks off like him too. 'Tis the voice of the titlark courting. The spring migrants have come out of the foggy sea, and filled the great gaps left by autumn migration, and cold, and birds of prey; for your hawk and harrier loves a tit-lark as we do a snipe. All through the hard grey winter have the resident titlarks been seen by the melting ice-fields, pecking hungrily at the worms and insects thrown up by the thaw, or feeding round the fenmen's mills. But now their troubles are over, and they are disporting their little bodies before their still more sober-hued mates. Up goes the cock again from his reed-bed or marsh, his little body swelling with his simple love-song, then down he comes, when his appointed course is run, into the grassy marsh.

Already it is building its compact little nest of coarse grass, fine grass, and little pieces of rush, building it securely in the moss on the marsh bottom in a sheltered spot beneath a tussock of beaten-down grass or rush—that serves it for a roof—a covering, however, which does not protect it from the visits of the ubiquitous cuckoo, who is already chasing, or being chased, by some of its tribe across the grassy seas—such is the mysterious relationship between these two birds.

But let us find a nest, for there flies a bird, shutting his wings and rising into the air and falling back again, reminding one of a swimmer cleaving the glassy waves of the sea.

There he goes across the wall into the soft marsh. Now he hovers for a moment over a spot, and dips into the marsh with something in his bill—he is feeding his mate, for 'tis too early yet for young birds.

Jumping the sleeping water in the dike, we beat about the soft marsh, putting up a bright-eyed peacock-butterfly ("King George" the fenmen call it), and several brown and yellow moths, all of which flying creatures delight his little palate, for he is a lover of dainties. Ah! we have put up the little pair; let us follow them to the dike-side.

Watch! he has got a little worm, and they are walking away up the dike beyond the kingcups. Let us follow; but hold! they are looking at us. On they go past that patch of cuckoo-flowers—he is going to feed the hen. No; they move off again. Let us follow; they bid us follow towards the mill. On we go past an old thistle-stalk; the saucy pair eye us askance, and still he holds the worm in his bill. They stop and seem to confer, yet she eats nothing. By jove! they're fooling us, leading us away from the nest, and the big fenman cracks his sides with laughter and shouts, "The little warmin." Yes, they took us in, as they have done before by feigning lameness; and yet he feeds his hen whilst she sits, so we must go and hide behind the wall and wait till they return to the nest; for that is the easiest way to find the reddish eggs—an you wish.

But hold! ere we have hidden another appears, looking lighter in plumage than he did in the dark days of winter, when he resembled the hedge-sparrow and starling in the white snow-fields. Watch him; he is hunting along the bank, the colour of his feathers being a shield as he hunts in and out of the grassy tussocks. Naturally shy, he sneaks along something like a rail in a reed-bed, taking advantage of any bit of cover, like the skirmisher he is. Look! he has sighted his prey yonder, a yard ahead of him. You see he is running swiftly towards it; lo! he has seized and swallowed it, and look! he has already begun skirmishing again. But he has stopped again at yon little silver pool and is sipping the water; and now look how he preens his feathers, raising his head as suspiciously as a guilty lover. He is in *déshabille* now, his wings hang loosely by his side, his feathers are shot out; still his quick little eye sees an insect drop into the pool, and, after a swift glance all round, he seizes it quickly, looks round swiftly again, and darts his head down into the feathers between his legs for another insect that irritates him. But up rises the head again. There! you moved, and he is gone like a startled hare, and we will follow; for we have disturbed the breeding birds to-day, and they will sit about on the tops of blossoming gorse-sprays, budding brambles, and even on the shaft of that old fork left idly in the marsh, whence they will fly along the heaps of stuff till they think they have wearied you, notwithstanding their eggs are cooling.

All summer long, from May until the golden harvest, you may find their eggs or nestlings over the marshlands. There you may see them hovering with laden bills over their cradles or flying up in courtship, or with their young following the marsh-mowers, snapping at the moths and

butterflies so dear to them; or else they are feeding a
voracious cuckoo, and following him over the marshes, pan-
dering to the great silent bird through August, until he
ungratefully goes across the seas when the guns begin to
boom across the stubble, leaving his foster-parents to fly
aimlessly to and fro looking for their lost child; or mayhap
when the ever-new springtide returns, they give chase to the
first cuckoo, stung by affection, and thinking perhaps to find
their dear foster-child, for whom they so willingly sacrifice
their own flesh and blood. Strange, insignificant, mysterious
little birds! what is the riddle, what the mysterious love
you bear the great selfish chortling cuckoo? What the love
and enmity that by turn stirs your soft souls to deeds of war
or loving self-sacrifice?

CHAPTER XXVIII

RED-BACKED SHRIKE

THE dark leafless trees and pale marsh crops of the winter landscape have gradually melted into a green screen that rises from a green sea. Indeed, April is here, borne on mild yellowish wings—April, the month of the blue cuckoo and the red-backed shrike—for they often come over the water plains together. Indeed, I have heard the red-backed shrike called the "cuckoo's mate," though wrongly, that name being reserved for the "wry-neck," which is rare on the marshlands, where the red-backed bird is not unfrequent, as you may prove for yourself; for he is often to be seen in early spring with his slow lapping, fulfar-like flight going across the marshes, uttering his whistling note as he makes his journey through the blue to the old hedgerow by the green loke leading up from the marshes. He knows it well, for he has nested there for several years in succession, and there has been no one to disturb him. See how he alights on the familiar old hawthorn "sprag," as the fenmen call a spray. Perchance a brother has sought his old nesting home in the cool retreat of an osier carr—a home rich with insect food for the expected family.

As you go along the loke, gay with yellow broom, you see the pair sitting upon the bending "sprags" against the blue —for they never alight or sit in a hedgerow, but always upon some outstretching "sprag," whence they can view the flat wastes around them.

Up they fly and away they go down the loke to that out-lying "sprag" of bramble; and if you "hide up" in the hedge you will see one hunt mayhap, for the evening shades are calling forth the moths, and droning beetles begin to fly and drum; and as you watch him, suddenly a little brown moth may come down the road, and you see his crest rise, you hear him utter a peculiar call—*whorrt*—and, darting from his look-out, capture the soft-winged little thing in his hard bill, returning to his perch to devour him. Moths and beetles and flies are his favourite food, and he always captures them in this way all through the summer.

When the end of May has clothed the hedgerows with verdure, the butcher-birds begin to weave their large slovenly nests of old grass, roots, moss, wool, and sometimes horse-hair; placing them either in a thorn bush, an osier, a bramble, or even in the hole of a tree—one once built in an old apple-tree I knew well—the hen there laying five eggs; but more often six.

The old birds never go far from the nest, and you may see them perched aloft on a rocking "sprag" at eventide, or in the day hunting for their prey, which, unconscious of them, swims by in the airy stream before them.

And when the nestlings leave the nest, you may see them huddled five or six together on a short branchlet, the old birds feeding them, and the cock calling from time to time with a voice like that of a small hawk. The cock is the chief caterer to the happy family; and should you capture one and place it on the ground, and bring your finger up to its bill, it will hop backward; and should you for fun put him into a yellow fulfar's nest close by, he will unceremoniously bundle his hosts out. And if at such season a cuckoo pass that way, the cock will dart boldly at him and drive him off. Indeed, he will turn birds from their nesting quarters if he have a mind, and cares not one jot for any intruder when there are young in the nest.

The great grey shrike I have seen once or twice in Nor-folk, but it was a brief vision—a momentary hovering like a kestrel and a jay-like flight into the winter greyness, just as one sees a bright meteorite on a clear November sky.

CHAPTER XXIX

THE SPOTTED FLY-CATCHER

THE irises are high in the dikes when the fly-catchers come over the seas to build their mossy nests of moss, cobwebs, horse-hair, and wool in the thick ivy climbing round the old elm-trees surrounding the fenman's garden ; nor do they ever leave their nest far. I have heard them early in April, but the end of the month is a surer date for them to appear ; nor are they long about the little gardens abutting on the Broads. Often they begin to build their nests in the cosy corner between an elm sprig and the mossy, green, ivy-clad trunk—green with the rain-paths of years—paths where the soft water has run down from the spreading topmost branches for many a season.

Both these serious-looking sober little birds build the nest, and both sit by turns. From incubation-time all day long through the breeding season you will see his speckled little bosom flitting about the elm branches, or else sitting like a red shrike upon an elm sprig, he watches with his big solemn eyes for flies and moths flying in and out of the ivy, and suddenly darts upon them as does a bird of prey, seizing them and returning to his perch to look for more.

Indeed, I think he builds in the ivy round the elms because in the long spring and summer days this cool green winding-sheet of the elm swarms with moths and flies, and the young and old fly-catchers find they will not have far to seek for their dinner. And in summer you may hear them singing their sweet wren-like song, but more especially in the thick sultry noontide or afternoon, when not a breath stirs the gorgeous yellow irises in the green dikes, and never a reed tosses its rich black tassel beneath the hot blue sky.

Listen, and you may every half-hour, sure as the castle-watchman's call, hear him burst into song for a short period, and cease again to hawk for flies until his watch is up, and the song is given forth again; or you may hear him calling *egyp, egyp,* as he feeds his brood, be it the first or second family, for he builds twice if robbed.

In sooth, the voices of the buntings, whitethroats, and fly-catchers may be called the voices of the summer-tide, for when they grow silent we know that chill autumn and dreary winter are nigh—know that the iris seed-cases are bursting in the dikes, and the pike seeks the deep water. And go if you can, with the fly-catcher to some sunny clime, where, in the thick hot noontide, you may there hear his sweet little song, and forget that it is winter anywhere.

CHAPTER XXX

THE SWALLOWS

ALL four of the "swallows" (as the fenmen call swallows and martins indiscriminately) frequent the Broadland, and by their form and flight and gentle twitterings add life and grace and elegance to the Broads and marshlands.

The first to come over is the swallow, a few stragglers appearing early in April, each bright bird's arrival being passed on from mouth to mouth, for he is an earnest of spring, a promise of flowers and love and hawthorn-sprinkled lanes. He is the most patent outward and visible sign of coming spring, though the seer has been expecting him for some days; for has he not noticed the midges dancing over the dikes and sappy marsh plants, and the swallow never comes before they appear. It is remarkable how regularly a few swallows and house-martins arrive, and then follows a lull of a fortnight before the main body come over, and begin hawking for flies over the still waters of the lagoons, whence the lily leaves are rising and the young reed-cases peeping forth. And often after his arrival a cold wave sweeps over the face of the land, filling the thrushes' nests with snow and sleet, cutting down the flies, and so starving the swallows, whom you can find lying dead by the roads and dikes, or see flying numbly a few feet over the hard resounding rimy marsh.

And daily in early spring, as you sail through these quiet waters, you will see the handsome swallow flying over the water in search of midges; and as the sun gains strength and insect life increases, they leave the rivers and hawk over the marsh grasses for moths, or along the white roads.

And towards the month of May you see pairs flying through the air twittering—courting—ere they begin to build their daub-and-wattle cradles beneath the bridges across the dikes, or under the bridges over the water-mill outlets, secure from wet. You may watch both birds in the clear, bland air, carrying bits of dead rush, straw, and soft pellets of ooze, plastering the nest together in primitive form; and as you watch the tame birds flying to and fro, making a sound like the shutting and opening of a fan, by his smaller size and greater blueness you may know the cock. All through hawthorn-blossomed June and into July they are nursing—for they raise two broods—on the pleasant marsh; or later they take to old sheds on the sandhills, for flies are plentiful there; and their five spotted eggs are laid in a curious bed of feathers, and the final sitting begins, both birds taking turns at incubation. Whilst one is sitting the other is abroad feeding on moths startled by the mowers from the coarse swathes of marsh grass, on flies hovering over the reed-beds, on dead flies blown into the water and washed to leeward by the wind, or upon flies darting over the hot sand-hills. At these seasons, too, when the young are hatched, you may watch them hawking all day, often attacking insects too large for their wide throats—victims they are forced to let escape after all; or flying low over the water, dipping into it, and at times dipping so deeply that they cannot get out and are drowned; or gathering "blight" from the reed-beds—or you may hear their little *pheet-a-pheets* as late as half-past nine on a summer's night as they hunt along the dozing trees and hedgerows.

And the young remain in the nest until they are good fliers, when they adventure the unknown; but not altogether, for some leave the nest sooner than others, and at this season you may see four or five young swallows sitting together on a bramble or old rail. Still the old birds hunt for them, and you see the young fly up to meet their parents, taking the food from them in the air amid much twittering, returning to their perches, where they wait in solemn silence whilst the parents hunt about, filling their crops with flies. And so from daylight till late at night they eat and grow, and by the time the equinoctial gales begin to blow they are prepared for the sea-voyage, and in September you may see hundreds of them on the lee-side of a house or barn waiting for a suitable wind ere they launch upon their journey, often flying straight up into the cool, clear air ere they steer for foreign shores. Some old fenmen declare the young birds leave a fortnight before the old birds, but I cannot tell. But after the main migration some of the young linger. I have seen swallows every month of the year except January and February, but I think these lingerers often pay dearly for their tardiness by death; but I should not be surprised to hear that some "hide up," and remain over winter in this country. Indeed, I am inclined to think they do.

The swallow is not such a lover of the water as the sand-martin; indeed, after early spring he does not hunt so much over the water, for flies are more plentiful inland.

If his nest be robbed, he will build a second and a third, but as a rule he is content with two broods a year.

The bird's flight seems ever a joy; he seems to contract his wings with delight, giving a few quick beats and leaping into the air, which he seems to embrace with passion.

As a decorative bird the swallow is one of the most charming, and I know of no more joyous vision than to watch a little flock of freshly-arrived swallows with their brilliant metallic blues and greens, blacks, whites, and chestnuts, describing beautiful figures over a still lagoon as they hawk for flies, and filling the soft genial spring air with soft musical *pheet-a-pheets, pheet-a-pheets*, now dipping into the warm waters, now darting this way and that; visions of joyous and beauteous life one can sit and behold all through the spring day. One wants nothing when thus occupied, and the world is young and green and full of promise, a deep contentment possessing the beholder's mind. Of all dancers, the swallows, the aërial dancers, are the most graceful, the loveliest.   Yet men and cats kill them, the one for " sport " or to decorate their womenkind's hats, the other for " sport " too, for they do not eat them.

CHAPTER XXXI

THE MARTINS

THE house-martin is rarer in the Broadland than the swallow, but his white rump is frequently to be seen some three weeks later than the swallow's burnished colours, hawking over the green marshes, calling with his soft *preet-a-preet*. But this brief period of fly-hawking lasts but a fortnight, when they break into pairs, and seek the beam eaves of some shed or mill or the rafters of some high-level wood bridge; for, unlike the swallow, they seem to prefer higher and loftier nesting-places; for which reason perhaps their nest is more protected, being cup-shaped with a small port-hole for ingress and egress; and in that rude chamber the white eggs are placed on feathers. But there is often considerable noise and fighting before the pairing is satisfactorily settled, and even then they have to fight that thief, the cock-sparrow, who sometimes quietly appropriates their nest and ejects them. Altogether they seem more pugnacious than the swallow, as you may see when one has captured a moth too large for his wide mouth; for immediately another will be on his track and try to seize it, and what a *preeting* there is then. But they do not remain long on the marshes; as I have said before, they go up to the houses to build their coarse cradles of grass and mud, which they collect from the dikes, ponds, and roadsides, preferring always a dark ooze of a clayey nature, to a yellow loam. And they are wise, for the nests made of this ooze do not crack so readily with the heat, and last for years, the birds often returning season after season to the same nest. Indeed, the swallow uses the same ooze, and he too returns to his nest year after year

if it be undisturbed whilst he is rearing his family; but if the
nests be broken down after they are gone, they will all commence building on the ruins of the old house.   The house-
martin is more persevering than the swallow, and if an urchin
take it into his head to knock the half-finished nest down, it
will build it again and again unto the fourth or fifth time.

And in this cosy warm bed of feathers—for the swallow's
bed is scantier and open to the air, for he is hardier—the
martin often mixes horse-hair and grass freely with layers
of feathers and wool.   And both sit, and the young are fed
in the same manner as the swallow.   When they are sufficiently strong, they crowd, as do the swallows, to the edge
of the nest, and the old birds hang upon the nest and feed
them so.   But the martins remain in the nest till they are
very strong on the wing; indeed they often return to their
nests after their first flights from the cradle.   This short
flight in ascending curves with quickly beating wings leads
them to a height whence they seem to swoop down a curve,
and so fly on up and down.   And they never leave their
birthplace far, hovering round about their cradle, being fed
on the wing until the golden harvest-days are over, when
they collect into smaller flocks than do the swallows, and
generally leave the country before them; for they are not so
hardy, as I have said, and they are wise to flee the chill grey
fogs of England whilst their bodies are still strong.

THE SAND-MARTIN.

The greatest lover of the Broads is the hardy little sand-
martin.   He is often the first to arrive; a swallow or sand-
martin heralding the coming spring in turn, and when he
does come, he is nearly always to be seen by the water-side
sitting on old gladen stalks, gracefully resting on a bent
reed, or hawking over the rippling broad for flies—flies that
are born and bred in the reed-beds.   But the flocks soon
break up, and seek some sand or gravel-pit or the clay-pits

in brickyards on the edge of the marshland, returning year after year to the same holes. These dark tunnels they have burrowed deep into the bank, where they place their floating basket-like nest of fowls' feathers, upon which the milk-white eggs, known to every schoolboy, are placed and the young hatched, and where the young are fed on flies and moths till they are pretty strong on the wing, when they go in a body to the Broad, and all day long from dawn to sundown you may see them skimming just over the reed during rain or thunder-squall. In the still oppressive noontide or at dusk or early dawn, it is always the same, there they are always hawking and twittering, the old feeding the young on the wing, and consuming enormous numbers of flies; for these birds are far more numerous over the Broadlands than either the swallow or house-martin, as may be seen when they congregate in large flocks of thousands, sitting by the sounding sea, or lining miles of bending telegraph wire, or covering the reed or gladen beds preparatory to starting forth; and as they fly up, on the start, darkening the air, it is difficult to conceive how they do not dash against each other, so black is the flock.

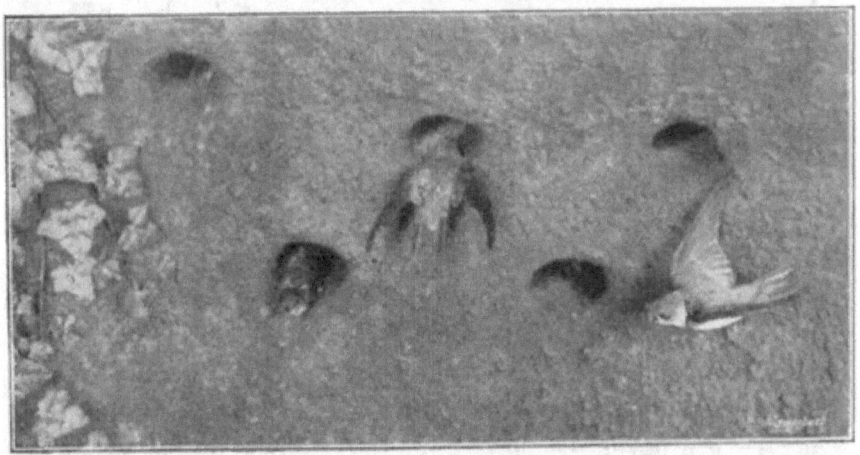

SAND-MARTINS AND BURROWS.

CHAPTER XXXII

THE GREENFINCH

OR "greenulf," as the Broadsmen call him, is the dirtiest bird alive—he fouls his own nest, always a stupid achievement. But you shall see; for this is a beautiful April day, and the bland new leaves of the hawthorn have covered the gnarled stems; for the greenfinch is wise enough not to build till the thorny skeletons are clothed with verdure, and oftener with beautiful "may," that perfect flower, that catches the eye of many a courting greenfinch.

Let us walk under the blue sky, flecked with soft, snowy cumuli, passing some geese sleeping upon a rushy marsh, the sentinel on his grassy hill calling doubtfully as we pass up the white road towards the village, whence comes the fish-hawker calling through the green hedgerows, sweet with the voices of the whitethroats and yellow-buntings. As we near the village, the cottagers, in tucked-up dresses, pop into their doors with mops and buckets, peeping shyly from their diamond-paned windows at the "foreigners." We are overtaken by a heavy field-cart as it jolts along the village street, the driver sitting sideways, his hob-nailed boots dangling down against the rich bay skin of his strong horse, and as he leaves one end of the village we leave the other for the haunt of the greenfinch—a grassy loke bordered by very tall whitethorn hedges—a loke leading down to the marshlands.

Directly we enter the cool retreat, so fresh and bright in its greenery, we hear the sweet voice of a yellow-hammer, and also the chuckling, loud and shriller linnet-

like calls of the greenfinches, and as we go along the loke
a darker patch of green, calls with the voice of a sedge-
warbler across the light green background, 'tis the "greenulf"
himself, and this is greenfinch-land; here the hardy green
bird loves to build his green mossy nest in the crooked fork
of the whitethorn some few feet above the bank. And he,
like the starling, is careless of the prowling village cats, for
they love the flesh of neither of these dirty birds.

Should the winter linger late in the land, however, he
chooses a glossy prickly "hulver" tree or the intricate recesses of ivy bushes, for cover he must have, but the cover
of the bland young hawthorn leaves is his favourite dwelling-
place; and mayhap he may wait upon the lingering spring,
for I have seen their nests when the dipping mowers were
felling the golden corn on the uplands; but more likely
these were the last feeble efforts of a many-time robbed
couple, for the speckled white eggs in the mossy nest are
dear to the village youth; but the custom of raising two
broods a year, numbering four or five in a brood, remains.

Now let us work into their home, blue with the delicate
eyes of the speedwell, and watch. Ha! the fluttering and
chattering! it is a fight, a fierce fight between two cocks,
who dart at each other with their powerful bills, more
powerful than that of the house-sparrow; and if you will
but let him try, either of these cocks will raise a black blister
on your finger with his weapon. There they go with drooping wings and tails erect; their feathers bristling; away
they go down into that thorn by the yellow broom. Look!
here comes another, his bill full of seed for his young; he
is early, but then the season is early, and he follows it.
He has just come from that lonely piece there. Let us
watch him. There he goes into that tall well-grown whitethorn, and there is his green mossy nest. Climb up and
see. Oh! there he goes back, flying as if he had lost all
his strength, to decoy you away from his young; but as you
look into the horse-hair nest, the edges dirty with dung and

scales, you learn that the birds are of great size, for the old
birds carry the sacs of meconium away until the progeny
is large, when they dung where they list—dirty birds that
they are. Now seize one. As you hold him screaming in
your hand, all the rest have blundered out of the filthy nest
and run away, for they cannot fly; and indeed it is their
custom to leave the nest before they fly, so it is no startling
innovation.

And if the naturalist hunt the hedges well, he may find
a dozen nests, as I have done, in a short hundred yards; but
the young are best left alone, for they are useless as songsters,
though the farmer likes them to feed his ferrets, for he has
scarce a greater enemy than the greenfinch. And if you
watch, you will know why. Any day in summer, if you
frequent greenfinch-land enough, you may suddenly come
upon a chattering crush of birds sitting on some long thin
branch, eagerly reaching their open mouths to another bird
sitting on a higher branch. 'Tis the greenfinch feeding her
nestlings upon the farmer's turnip-seed. And such a commotion they make in the tree-tops, all greedily fighting for
a turn at the mother's crop, feeding upon the turnip-seed
mash like pigs. Then go you to the turnip-seed patch, that
feeding-ground of the linnet tribe, and mark the flocks upon
flocks of them feeding ravenously upon the farmer's ripe
seed—their favourite dish—and rejoice that their end is
often to fill a ferret's long lean stomach.

Later on, you may see flocks of them with other thieves
in the wheat-fields—pairs here and there fighting fiercely
for their food, for they are pugnacious, quarrelsome birds,
especially in the hard, black days of winter, when food is
scarce, and they scour the country-side with chaffinches,
bramblings, and buntings, hens and cocks all together, as is
the custom of the finches and buntings, all hunting over the
stubble and " new-lays " for food. At such season, so intent
are they on fighting, that they will allow you to approach
within a yard of them.

Indeed, they are a constant terror to the farmer, for from his turnip-seed they go to his corn, and thence to his bean-fields, champing up the ripe beans with their powerful bills. They care not whether the beans lie open on the black plants or be left scattered near the farm-house by the harvester—they are sure to get them. And when they have champed the last bean, they flock to the lonely stacks, turning them green with their glossy bodies, covering the rich, lovely heap with a murrain evil as locusts; but they are more easily affrighted, for many a shot is fired into the living green and a score or a score and a half of mangled green balls are left behind, food for ferrets. Still the nuisances return again and again, until mayhap the flock is thinned, when they often fall victims to the wide-mouthed sparrow-nets. And then comes the pinch in dead winter, when the grain is all housed and eaten. Then in the pale sunny mornings of early winter you may see flocks of them drop softly from the blue grey sky into the decorative alders by the cold riverside—they together with the long-tailed tits of sporting flight, and the garrulous redpolls, whose little wings embrace the cold air as they fly more quickly in short curves from alder to alder. But the company of the tits, redpolls, and pick-cheeses is deserted as soon as the barley is sown, for they are off after the grain; and once again the farmer curses their green limbs as he sees them running over the glistening clods like living plants, so green is their plumage in the winter sunshine. And such is the greenfinch, a dirty bird in his youth, yet loving a quiet paradise to court and build in, and a bold, quarrelsome thief in his manhood. And his end is to feed the rat-catcher's ferrets.

CHAPTER XXXIII

THE HAWFINCH OR COBBLE-BIRD *

WHEN the yellow leaves have dropped like great pale dead butterflies through the low grey skies of autumn, and the canker-riddled cauliflowers and budding sprouts are eaten up, and the millman's garden is a slippery morass of decaying vegetation, and dripping branches lie naked to the sky, the hawfinch deigns to visit the Broad district at rare intervals, and he is generally to be found in a deserted garden beneath a bare-branched, moss-stained bullace-tree, for he loves to crack the hard black cobbles with his strong bill as dearly as any schoolboy.

Mere glimpses of this handsome bird against a grey background is all we get in the Broadland, brief glances scarce worth recording, and the proud-looking bird with the curly feather is gone, so we know little of him and his ways.

* I have seen but one living specimen of this bird in Norfolk, though two stuffed birds I know of—one shot at Reedham in 1890, and one at Hickling, 1891—each one with bullace seeds in their mouth. In Surrey I have, as a schoolboy, frequently found the hawfinch's nest.

CHAPTER XXXIV

THE DRAW-WATER

THE goldfinch or "draw-water" is not a bird of graceful build nor sweet song, yet is he dear to the Philistine who loves variegated colours, because he satisfies a rude barbaric taste for colour; for he is a "gay bird," and he is great at parlour tricks, like his lover; for cannot he draw his water and seed to his cage by a simple mechanical contrivance? And so he delights the populace, as do the performing elephant and the contortionist.

And of our cage-birds he seems most ill at ease, and is perpetually rushing from one side of the cage to the other, and if he be given half a chance he will escape, for he is quick and a swift flyer, and returns to the marsh, where in sooth he is seen at his best; for at a distance, flitting restlessly with quick jumps from thistle to gorse-bush in the bright sunshine, he delights the eye, for 'tis an ever-shifting ball of colour flitting over the sere marsh-crops; but when you come to regard him in a cage, you find him ill-shapen, restless, bad-tempered, an indifferent musician, a mountebank and imitator, and a lover of rude noises, for he sings never so sweetly as when a millman's engine is rattling, pumping forth the marsh water into the rivers.

However, he has taste when building his nursery, for he generally chooses a fruit-tree, preferably an apple, covered with madder-tinged blossom; for though he pairs very early in spring, he does not begin to build till the middle of May. In some mossy crook he builds his neat small cradle of moss, and he is a good husband, taking his turn at sitting and feeding the young with flies and maggots.

Though shy birds, if robbed of their young they are very bold, and I have known them go regularly into a cottage to feed the captive young in a cage; but the cage was moved by degrees from the nesting-tree to the cottage table—an interval of a day elapsing between each stage. On the other hand, if captured when old, they will often sulk to death, or "die of skulking," as the fenmen say.

There is a superstition amongst cottagers, that if the young birds die, the old birds have poisoned them; but the mystery is generally to be explained by looking into the seed-dish, where pure hemp will be found—a seed fatal to young birds.

In autumn they collect in small flocks, and may be seen beating the thistly marshes or flashing over the snow in mid-winter, when they look at their best, and indeed are then very tame, and can almost be taken by the hand.

But they are becoming rare in the Broad district, and though they may be seen on the marshlands and in the elms and cars by the river-side, it is not an everyday picture.

The sentimentalist, whose heart is often better than his head, often raises an outcry against caging birds; but if the young bird is taken from the nest before it recognises its parents, there is no cruelty in the matter, for they never know the doubtful sweets and dangers of bird liberty, and will at times, if they escape, return of their own free will to their "prison." Should you wish to take young birds for the cage, you must watch your brood every day, and so long as these formless creatures upon your appearance stretch forth their ugly necks and open their mouths for food on your approach, so long they are ignorant of their parents, for 'tis a merely reflex action. For when they begin "to take notice," the ugly maw is no longer opened at your approach. You must not, however, wait for that period of development, but take them just before, when they are fledged. Seize them boldly, forgetful of the sentimentalist, and cage them, placing the cage on a stout branch near the nest, and the old birds will feed them.

A few hours later you may move them up the whitethorn decorated lane for a couple of hundred yards, and so on up to half a mile, but no farther, in one day—that is a young goldfinch's infant day's journey. So by degrees you decoy the parents to your garden. When the parents get tired of feeding your captives, make pills of eggs and flour, and give them together with plantain-seeds and thistle-tops, and so you shall educate them to sing when you blow your slender fire with the bellows; and, if you be a woman, destitute of human lovers, you may teach the brown-billed bird to kiss you; but a man's kiss is preferable.

Unless you wish him to sing another's song, for he is of the mean tribe of plagiarists, keep him when young out of hearing of other birds, as he, like many a human parrot, prefers the songs of others to his own. But for any purpose whatsoever, I do not think his company is worth his keep.

CHAPTER XXXV

THE SISKIN

Is not a bird of the fens, though in spring his song there recalls the sedge-warbler, and his peculiar little call, like that of a miniature guinea-fowl, may be heard among the plantings. But he never stays by mere or river to cradle his young, and 'tis a pity; for a dearer, coyer little fellow never lived. Bright and quick as are most insect-eaters, he is one of the most attractive of pets. Running over the bars of his cage, his little sedge-warbler-like head darting this way and that, creeping about his quarters like a mouse, yet affectionate withal; stopping to eat seed from your finger, and anon playing at hide-and-seek through the nesting cubicles, then diving down to his pond and chirping so prettily ere he begins again to flirt with you, as first he would and then he wouldn't—for he is typical of lovely woman—willing, yet unwilling; giving, and yet refusing; leading you on, then throwing you off; affectionate one moment, inconstant another, he is the very maiden of birds, and the most desirable to keep by you.

CHAPTER XXXVI

THE COMMON SPARROW

A PLAIN little burgher, greedy, pugnacious, and harmful, and would that all sparrows had one head, and that I might be allowed to silence for ever the infernal chattering of that commonplace pest.

He is in no hurry to nest—often waiting for that artist, the house-martin, to finish his cosy cradle, when, with his strong predatory beak, he will drive the beautiful bird from his house, and you see his devilish leer as he peeps forth from his stolen shell, and prepares to perpetuate his under-bred progeny. Indeed, do you but try conclusions with him yourself, and he will try his sparrow-best to nip a piece out of your finger; for he is vengeful, like the low type he is. Indeed, I know of a case where a rat-catcher sent a puppy to retrieve a winged sparrow, and the little beast bit the puppy's lip, and would not let go, so that he ran back to his master, who took the bird away from his hold, when the little vixen fixed on his finger, and he had to crush its useless head to make it let go.

When he does not rob an artist of his nest, he will pile some straw carelessly in a rain-spout, forgetful of the thunderstorm!—under the eaves, or in a hole made by pulling out long reeds from the thatch, where the hen will lay her speckled eggs and raise her harsh-looking youngsters. And so much is she in love with her progeny, that she at times begins to lay fresh eggs—in all, six—whilst the young are still living. You see she is a willing conquest, and fond of her lover's embraces, though she does pretend to purify herself with dust-baths. At times they prefer to make an unsightly nest in the ivy on an old elm-tree, but this is not a common practice.

In the early morning the garden is filled with their infernal chatter, and they are busy pulling off buds or splitting them open with their hard bills, or else they are robbing the green peas, or plucking off the tender tips of the young pea plants, or pulling the primroses on your lawn to pieces, or playing general havoc, and drowning the sweet soft voices of the insectivorous birds with their senseless chatter, as well as frightening other birds from your garden. Then the foolish young birds appear, for there is scarce a greater fool than a young sparrow; he must fly where there is danger, and if in any trouble he has no idea how to extricate himself. I one day watched some young birds busily eating newly-born ants—the winged progeny—for they won't touch the others. One of these ants fixed on to a young bird's foot, and instead of eating him off, as a siskin would have done, he flew away shrieking, helpless as a soldier in a shipwreck.

And as you succeed in sowing your seed, the sparrow is there for an early feast—to pick holes into the ridges, to eat wholesale until the corn ripens in the fields, when your pest deserts you for the farmer, whom he injures most of all; for though he spends money early in the season on bird-catchers, who "scrap" places and put out oats to "coy" them with; still, when the corn gets "for'ard," the sparrows go to rob the fields. I have seen hundreds of these clumsy, commonplace thieves amongst the ripe wheat and barley, eating the grain planted along the fences; and they collect in greater numbers where the hedgerows surrounding the fields are high. But the sparrow is not proud; he will eat hog-wash, and is very fond of "fare-a-faking" (the cottagers' pigs' straw), as well as playing havoc in an early spring onion-bed, having as keen a bill for a young Banbury onion or early shorthorn carrot as a cook; nor will he always wait for the corn to ripen, for he, with the greenfinches and ringdoves, is fond of picking off the young corn when just up, at times even building amongst the corn and turnips to be nearer his food; so the marsh farmers say, "He have a lord who go ahead just as we do." To give him his due, I have seen him eat "green-fly" and cabbage white caterpillars; but there was nothing else to be had. In summer-time the sparrows often roost under the hedgerows to be nearer their food; but when they have done gleaning, they draw up to the farm buildings and roost in the barn or in the stacks they are despoiling, and then is the farmer often avenged, and the sparrow-catcher comes of a night and shines his bright lantern, and the foolish birds fly at it like moths at a candle, and are caught in the clap-nets or sieves fixed to poles; and when the bird-catcher goes on to the sheds and barns and discovers his light, they fly at it with a rush—it's yellow, like corn, you see—and fall blundering on to the floor, an easy prey. There is no end to their foolishness. They will, when engrossed fighting for their ugly nests, come down a chimney all grimy and impudent; and if startled at night, they will fly straight into a room through an open window to the lamp airily as any moth. Is this creature of any use? Yes, he may be eaten. The breasts of the young bird stewed are good enough, as is a sparrow-pie and a jug of ale on a cold night. Moreover, he is one of the best birds for trap-shooting, for, unlike the greenfinch, he flies steadily and straight, and he is useful for feeding ferrets and captive merlins, who select his busy brains, however. But alas! his extermination is hopeless, for, like all low types of life, he is most fruitful, and, drat the bird! he, ay, even he, has his sentimental admirers—out of a pie! But they know nothing of art, natural history, gardening, or farming. And after all, is not England avenged on the United States for the War of Independence? She has cursed these States with the curse of the sparrow, to which the plagues of Egypt were as playthings.

CHAPTER XXXVII

THE TREE-SPARROW

OR "French sparrow," as the fenmen call him, is called the tree-sparrow, mayhap, because he has nothing to do with trees. I have never yet found his nest in any place but a shed close to a cornfield, though I have heard of his building in a hole in a tree.

These little birds are common enough in the Broad district, and any year I could go and get dozens of nests in flowery June, and he is easily distinguished from the common or fiendish sparrow by his chestnut crown, his lighter colour, and smaller size. His eggs, too, are smaller, darker, and prettier than those of the common sparrow, and they are laid later in the year than those of his *confrère*.

Though many of them are resident, some come across the North Sea in autumn to thieve and rob the farmer's fields; they revel in a "new-lay." At the latter end of May, when these pests seem to increase in numbers, they may be seen working in parties with the common or thieving sparrow. They are, however, shyer, quicker in their actions, and seemingly more intellectual. Nor are they so injurious; but they must be placed upon the condemned list, for the small good they do in eating grubs and caterpillars in no way pays for the corn they steal. The gardener need not trouble about them, for they do not seem to like to work too near man's habitations, although they will follow the plough with the rooks, peewits, and starlings, feeding upon the grubs and wire-worms turned up by the share.

Though plentiful in summer, and especially just before harvest, they seem to get scarcer in the autumn, and may probably leave the country, returning at the spring migration. And I would that he steered another course, and left old England on his port bow.

CHAPTER XXXVIII

THE SPINX

I REMEMBER one April walking through the tall willow saplings of a waterside coppice, sprayed with tender green shoots, when I saw two birds, brilliant in red, mauve, black, and white, whirr with swift flight through the moss-green stalks in the soft light, and alight on an ivied oak-tree, whence they poured forth their short sweet songs, ending abruptly as they flew off after the hen-bird in soberer dress ; for the gay-coloured chaffinches were courting, as I knew by the short sweet song and the *spinx, spinx,* of the hen-bird as she alighted on a branch and watched the amorous lovers contending through the coppice for her person.

I sank into a bed of moss, blue with hyacinths and bright with primroses, and, looking up the tall smooth green stalks, watched their love-flighting till the brightening sun had opened the water-lilies on the mere. Then I left the cool brake and sought the mere, and as I pushed from the low shore, I saw the hen-bird and one cock-bird flying off together to a large weeping willow that overhung the island in the lake, for year after year they have built in a heavy willow branch or in a shapely alder, placing their perfect little nest of lichen-embroidered moss, lined with the gay feathers of fowls, over the still waters, now blue, now silver, now rosy when the sun sinks over the reed tassels. And yet they are not happy, for mice dearly love their eggs, and woe to the spinx that builds in a hedgerow.

I remember another spring when I saw these gay birds tearing up my young radishes that flourished under the

delicate petals of peaches and nectarines in a south border,
when I meditated, " Of what use art thou, solemn birds, with
thy everlasting *spinx, spinx?*" And, truth to tell, 'tis a
characterless, disappointing bird, and never comes up to its
promise of spring. Though, in fairness, it must be said, I
have seen it gracefully hawking for flies from the ground like
a wagtail, whilst uttering little cries like a linnet. Only in
the glad May-time, when it is building its lovely little nest
in the whitethorn, now screened with new-born leaves,
and singing its short, sweet song, does it attract one;
for, after rearing the nestlings upon caterpillars green as
new-born leaves, seeds, maggots, midges, and moths—often
gathered under the haymaker's eyes—this bird seems to
settle down to a petty round of details—flocking together
with bramblings and greenfinches to rob the fields of newly-
sown corn until the hard weather drives them up to the
farmer's corn-stacks, where they often fall victims to their
lusts. Even in death they are useless. And yet they are
the most eager of birds to leave the nest, starting abroad
when their backs are thinly shielded with a few feathers—
long before they can fly. A restless, ever-attempting, never-
accomplishing bird is the spinx—a bird full of resolve, yet
ever suddenly halting in the heights of his endeavours: he
interests me not. And still they can cross the seas, for
flocks of them come over the North Sea in autumn with the
larks and rooks—"wheat-pickers," as the fenmen contemptuously call these bands of roving thieves—as if their two
children a year were not a sufficient supply of these dull
birds—the friend of the sparrow and dirty greenfinch—
the lover of the cage. For did not I once capture a
young chaffinch in my tool-house and imprison him, and,
finding his worthlessness, free him. But lo! at eventide he
returned to the tool-house, and allowed me to catch him and
replace him in "prison," where he seemed happy and contented. But his eternal *spinx, spinx,* so worried me that at
last I sent him away, and may the season have mercy on his

feathers, and may his detestable *spinx, spinx,* never resound
in my garden again.   Poor lifeless chaffinch ! whose energies
are exhausted in nest-building.   Poor unproductive dullard
who sings once a year and steals the rest of the season.
Hardy thou as the sparrow, uselessly and clumsily living by
thieving, grudging of song, dull in mien as a preacher, gayer
in winter, and gayest for a brief period in spring, thou lover
of winter bachelorhood, thou bird dear to the phlegmatic
German, good-bye!   Once a year, for a short season, you
silly nest-betraying male are tolerable, but for the rest of
the year—well, there is the sea, and dear Germany—go seek
it, and perish, *spinking* as you perish, you living chromo-
lithograph.

CHAFFINCH'S NEST AND EGGS.

CHAPTER XXXIX

THE BRAMBLING

IF the chaffinch be a living and vivid chromo-lithograph, the brambling is a dull, muddy-coloured example of the art, as you may see if you choose a grey day in autumn, and steal stealthily among the new-laid marshes near the roar of the sea. I say stealthily, because the brambling is one of the most alert birds to scent danger, and so is most careful of a useless carcase, like many another wastrel. As you approach the new-laid wheat-piece the many-sounding finches and buntings will be heard chattering, and the yellow loam will be covered with flocks of white, red, green, yellow, and brown, living flocks that seem to move swiftly over the face of the sea-marsh, regardless of the gunner lurking on the hollowed sea-dunes. If you watch them diligently with your glass, you will see the brambling finches, quick of movement, much resembling the chaffinch—cock resembling cock, and hen spinx resembling hen brambling. There, among the green-finches, linnets, yellow buntings, and land-buntings, you will see them feeding and moving restlessly about, and should you crack a rotten stick with your feet, up fly the cock bramblings, swift of wing, and the cloud of the finch tribe gleams like a rainbow in the sun as they fly off to a many-coloured fallow marsh, where the soil is filled with charlock seed, or " garlock " as the fenmen call it. Indeed, the charlock is the pest consecrate to the brambling finch, and since the charlock loves the marshes by the roaring sea, there you will find him from October to the last week in April, when, lo! he goeth whence he came, and the ripe crop of charlock blows unseen for him.

The fenmen cage him, nor do they find it difficult to tame him. Give hemp-seed with a free hand and he will soon grow reconciled to his new "prison," and favour you with his spinx-like call, or a more pretentious twittering—his own, but a poor thing, and not necessary to life.

But the vision of him in sober grey winter flitting over the embrowned marshlands, gives a tinge of colour to life, though it be crude and garish and impure; and as a forced bean is precious to the winter diner, so may the brambling be an addition to the dull wintry landscape. A chromo-lithograph may brighten a wall to keep a nightmare away, as it may produce one.

CHAPTER XL

THE "BLOOD" LINNET

AT the end of April linnet-land is sweet with the voices of nesting linnets—the new-comers from across the sea and the home-abiding birds—for they paired early in April when the king-cups were opening. The forest of flowering gorses standing by a dike-side yellow with ragwort is splashed with their droppings, the needley branches affording rafters for their pretty nests. As you push your way through the prickly gorse needles, clothes-tearing brambles, and long, dried, pale grass stalks, startling a feeding hedge-rabbit, who with ears laid back has been watching you ever since you entered the tiny forest, you startle these lively scarlet-flecked birds. A cock flies up with a piece of wool in his beak; a little further on another cock sits watchfully on a bramble spray against the loke, his bill full of flaxen gladen, down which he at once dips, or sings on, and commences sedulously to rub his bill against his shaking perch.

Sitting in the white grasses, grey as the hair of an old man, you get a peep of the russet marshes through a frame of decorative gorse—the full dike gleaming in the spring sunshine like the sun-embrowned face of a daughter of the marshes, and away there by the budding trees the black sprouting sallows resemble the fine pencilled markings on a shell found on this greenish shore of the marshland sea. As you sit silently in the cool undershadow you can hear the busy linnets coming and going, singing their flute-like notes as they dive off into the loke to search for building material. And if you watch patiently you may see many a pair go to

their unfinished nest—some mere woven foundations, for
they build from the bottom, weaving as they go, without any
layering or orderly strata of walls. As you watch, the air
grows musical with their mellow whistlings, chuckling songs,
and short sweet refrains—sweeter than the note of any
Æolian harp, yet as wild and free.

Tired of this panorama of dry frosted wood-work, for such
is a gorse wood seen from below, you rise and push through
the prickly wall, starting a whole colony of linnets from the
fresh green tops, driving them with cries away over the
marshes. Only one sits with great dignity on a furze spray
watching you intently—and he is grey, not in full plumage,
for thrice must the gorse bloom ere the nestlings don the
rose spots of manhood. And thus the rustic fenman believes
there are two kinds of linnet, the "hedge" or "grey" linnet
and the "blood" linnet; and yet these twain are one—as a
bearded man and smooth-faced athlete are one, and yet
twain.

As you hide again beneath a bramble bush, whence a
sluggish lurching hedge-sparrow has just been affrighted,
you will see that they are already beginning to return laden
with sheep's wool and gladen-spindle down, and perchance
one may fly into your bush, when he will suddenly turn
aside and pass off, just as a wayfarer hastily and unex-
pectedly met turns off with affected unconsciousness.

Thrice do they build if robbed, so strong is the instinct to
rear a brood; indeed, the gaping maws of young linnets may
be seen as late as harvest-time.

If you visit linnet-land regularly these lively bright little
fellows will grow friendlier, for they are naturally affectionate
and trusting, and you may see both birds sitting in turns at
one end of the gorse grove, whilst in another part some pairs
are building, and others again are feeding their young upon
mashed seed—turnip for preference—from their stuffed crops,
their bills thrust into the nestlings' gaping maws. For they
feed their young like the turtle-dove; and so loving is their

nature that both birds tend their nestlings, feeding them and carrying off the glistening little bags of meconium deliberately dropped by the cleanly young upon the grassy rims of the nest. For a dainty clean-living little bird is the linnet.

But there is another side to the picture. On a bright day in early summer you may perhaps be attracted by the glorious shimmering living yellow of a bed of turnip in flower, blazing beneath an azure cloud-flecked sky. Ah! that is a sight to live for each returning year.

But wait a little later, when the burning July sun shall have ripened the yellow florets into seed. Then you will see the linnets—hundreds of them—flocking to and fro amongst the strange-smelling crop; for not dearer is the heather, mignonette, or thyme to the bee than turnip-seed to the linnet. The farmer's man, with rusty old shoulder-gun, may stand sentinel by the precious crop and shoot them down a dozen at a shot at one end of the bed—still will they fly off with mellow pipings to the other end of the bed. At such season, by the gunner, you will see a pile of mangled birds—young and old if it be late in the season—food for his ferrets. And later, when the turnip crop shall have been harvested, you may see flocks of a hundred, ay, and two hundred, rise with chucklings as you approach the shorn seed-bed.

Later in the season, when the wild parsley-flower heads stream in unpremeditated grace across the marshlands, like a troop of laughing maidens tripping over the dew-loaded bents, you will find the rose-speckled flocks have gone to the osier cars in search of food. There in the sweet September evenings you may hear their song—a lullaby of the dying year. And as you stand on the white road, fringed with dewy grass, and listen, you may see the whole flock arise with chuckling laughs high into the air and spread like a fan over the sallows in the greying evening sky, whirling round and round beneath the misty stars ere they pitch headlong back into the osier cars, and their sweet voices

H

are stilled in sleep. And when the roaring equinox shall
have bared the osiers of their leaves many of them will start
across the salt sea for a warmer clime, whilst others, leaving
their old homes, will hie to the new-laid barley fields with a
greedy crowd of hungry finches, where they work mischief to
the husbandman; and none are more hateful to the barley-
sower than the green-finches and linnets, as none work more
evil to a newly-sprouted turnip field, except mayhap the
mischievous lark, who works much evil amongst the fine-
leaved turnips that have escaped the fly, but have still to
"apple."

And the bird-trapper knows this, and with his nets captures
hundreds, but your sweetest pet is always the young nestling,
reft from his home ere he begins to "take notice." He may
be even the happier in captivity, but old birds, who mind not
the net, never do so well—though they, too, are soon tamed.
And the old fenman knows this, and when he wants a "blood
linnet" for his lonely home he watches the nests in linnet-
land or on some low grassy islet on the marshes until the
nestlings are some ten days old, when he places a wicker
cage round the nest, leaving the old birds to feed their
young through the bars, for this they are willing to do, so
strong is their love for their progeny, till they be some weeks
old and fit to take to his cottage by the water-side.

But even such a sure method of capture has its cares, as
one old man told me who had trapped five nestlings. The
parents had fed them to a large size, for he was not over-
anxious to take them till the gorse bloomed by his cottage
door. Then he took them. And yet their capture was always
matter for regret to his simple soul, for one morning early,
soon after the larks had begun their matins, he heard the
penned linnets shrieking. Hastily rising from his bed, he
ran into the misty morn, and saw a bloody stoat, with hair
erect, clawing through the bars for the poor birds, most of
which he had already killed. Filled with ire, the old man
seized his yew-fir quant and struck at the stoat in fierce

rage, knocking him across a water-dyke, and breaking the cage to pieces, and allowing the only living bird to escape; so at one fell swoop he was robbed of all his pretty chickens. Nor was the stoat killed, as his noisome trail proved.

But **we** trust the old man has since captured a sweet musician who cheers his old age, and for a little seed and water sings to him year in and out. Dear, homely little minstrel! the poor fenman's chief love! Farewell, thou lover of the open air and the sunshine!

CHAPTER XLI

THE LESSER REDPOLL

WHEN the lush marshes are gay with purple and rosy water-grasses, waving beds of cotton-grass, and islets bright with ragged-robin, sweet-smelling patches of moist and many-twinkling orchid eyes, the redpoll's low song is to be heard in the islets of bramble and sallow that dot the green marsh-land sea—a song often heard as the bird flies by with its un-tiring linnet-like flight, a song that mingles beautifully with the insect-like voice of the grasshopper warbler, the speckled red-leg's cries, and the mournful peewit's notes. The rich marshes are growing ripe for the scythe ere the redpoll begins to nest, for he is a late nester — rarely building before the end of May. All the preliminary courting flights about the leafy sallow-bushes over, the hen bird begins to build that most exquisite little specimen of nest architecture —her rushy cradle securely lodged in a spray of sallow, or more rarely in a branchlet of alder, and most rarely in an apple-tree in some fenman's garden, or perhaps in a hedge-row ; but they love the marshland best. The hen with her little bill and feet deftly weaves the light shell of pale straw-coloured grass, then lines it most decoratively with a light bed of dindle fluff or catkin flowers, or, more beautifully still, with the sweet white down of the cotton-grass, or at times with a beautiful fringe of swan's down, the tips of the feathers only showing, and all curving beautifully inwards—the frail structure with its five blue eggs looking like some strange but lovely incurved flower of the marshes.

And if any fenman draw near her whilst engaged in this lovely work she and her gay little husband will flit about on the sallow branches, keeping close to his side, calling with their sweet linnet-like notes, for they sing more naturally when flying through the air in their peculiar chain of curves; and as the watcher draws off she will return to her work, and the cock may fly off high through the air like a linnet, flying a lovely pattern across the blue as he rises to the crest of each imaginary curve with a sharp snap of the wings, and descends another aërial curve, and so on till he and his mate are lost in the liquid blue. Presently he comes back shooting through the blue, and sweeps suddenly to the sacred sallow—the tree where he and his mate have reared their young for many a year past; for they always return to the same spot to nest if not disturbed. An old gunner had one pair build in an apple-tree in his little patch of garden, at a water-dike's edge, for two successive summers, and never would he allow them to be robbed, the "poor little warmin," as he said affectionately. And when the little blue eggs are hatched, they will never both leave the nest; for indeed, if you come upon them at the breeding season, be sure the nest is near by; they half betray it, the incautious little minstrels.

And in June and July you may see the cocks flying high over the green marshlands with food for their young. Very funny and serious they look as they fly with their curious flight across the sky. At such season, too, they fall victims to the farmer's gun, for they too love the turnip-seed, so dear to their relatives, the linnets and greenfinches; but whether they feed their young upon mashed seed I am uncertain.

And when the cares of nesting are over and the young can fly strongly, in the early autumn they congregate into flocks small and large, and descend across the glistening dunes where the grey-green marram gleams purple-tinged against the sky; for they love the chickweed that thrives thereabouts, and in your walks by the foaming waves you may put up flocks of them, sending them calling adown the wet sea-beaches, their short cries being lost in the roar of the everlasting ocean.

But when the cold snow-squalls come from out of the deep and the sea-rains fall cold and drenching, they flock back over the protecting dunes of sand, and seek the leafless alder-trees and eaves resounding with the gunner's shootings.

And all winter long, when the decorative alder-trees hang in the grey mist-world, you may see those trees covered with redpolls clustering all over their branches like bees ; and as the pale winter's sun gleams forth from a rift in the low sky their little bodies gleam fulvous in the still and brimming river amid the flickerings of the blue-tit, the spots of green-finch green, and mayhap the gayer winter reflections of the goldfinch, an arrangement in delicate colour surpassing the cunning of any hand to paint, a dainty pattern of delicate form and colour upon a silvery background; a decorative panel of exquisite beauty that sends a thrill of complete all-absorbing satisfaction through each atom of one's being; an exquisite, indescribable, evanescent poem for the sight, fleeting as the rose's perfume.   A worthy setting for such a perfect little artist as the dear wee redpoll.

LESSER REDPOLL'S NEST AND EGGS.

CHAPTER XLII

THE BULLFINCH OR BLOOD-ULF

THIS handsome bird is rare in the Broad district. Indeed, his home is the woodland, his delicious song being the embodiment of woodland music. Many a happy day have I spent beneath the cool shades of a coppice watching this splendid creature courting among the glossy holly leaves.

When seated beneath a cool "hulver" tree crushing the sky-blue hyacinths, you suddenly hear the low, soft call of the male, a short sweet whistle, to which the female responds in similar strain, though softer and lower, as becomes the woman.

At the sound the handsome cock grows eager and begins to walk up and down a green holly stick, his black cap, mouse-coloured back, and splendid crimson breast, feather tips, and tail lined with a ray of white below, flash in the sun as he walks with great dignity through the pretty glazed leaves, calling eagerly with his soft note. A rustle follows, and the plain-breasted little maiden trips coyly on to the same branch, calling softly, and she too begins to walk out towards the end of the branch and back like a flirting pigeon. And so they go on until suddenly they meet, their bills open—yes, they are actually kissing, and doubtless they seem to like it. Their kisses grow more eager; a hearty kiss and then a walk away from each other—and with such dignity! Indeed he is the embodiment of dignity.

A branch breaks and they fly away to separate trees, and then you shall hear such an antiphon as was never heard out of the woodland, for she is as good a singer as her lover. This antiphonal love-song is original, sweet, flute-like and unpremeditated, and, like the dove's song, seems to me to be the natural voice of the covert, the very essence of natural music, having the great qualities of freshness of variety and sweetness. And as he sings in a restrained manner, for very joy he at times flies up a few inches from his green seat and flaps his beautiful wings, resuming his courting walk amid the rustling leaves, finally lapsing into his sweet call-note.

In Mona's Isle they are very numerous, and generally build in the holly-tree, the hen being a very close sitter, so fond is she of the dear little spotted eggs. But with all his dignity and aristocratic looks, he is a sad fruit-thief and destroyer of buds, and his deadly enemy is the wise gardener; for he does nothing in return for his buds but sing, and the gardener is not swayed by such æsthetic emotions, so that I have known one boast of having killed five score of these beautiful birds in a season. That was in Anglesea.

Withal his strong parrot-like beak, he loves the strawberry, cherry, gooseberry, and grape. He is a tender bird and will not live long in confinement; but he is deservedly one of the familiar cage-pets, and to my thinking his song of beauty and dignified bearing make him the most desirable of all English birds. He is quite a character too. Forget to give him his green elm-shoots, or give him only a dish of canary-seed as a change, and he will sulk for a day, refusing the food with dignity, 'tis true, but still refusing it.

And what a termagant his wife is! She will chase him away from every cherry or strawberry thrust into the cage, gobbling it with true feminine selfishness for the good things of this life, and he bears it all meekly, even if she chases him from the bath, which he dearly loves, or sends him to roost below on a lower rung of the cage. He is long-suffering, but at times, when she goes too far, his crest rises, and his strong parrot bill makes a dig at her, and she is quiet, as feminine nature always is before force; for, like the faddist, she must be ruled, though not with too obtrusive a despotism.

In Norfolk he is rare, and rarely to be seen. For is he not the dignified retainer of the lordly house, the bright minstrel of the comely garden, and not the denizen of wild wastes and morasses; and as such I love him, and he is welcome to all the cherry-buds and fruit he can eat in my garden, even unto the muscats, which delight his soul and brighten his eye and sweeten his voice; dear lover of the sun-speckled plantations and garnished evergreens. He is the emblem of dignity, and we incongruously call him " Bullfinch !"

CHAPTER XLIII

THE LAND-BUNTING

THE buntings are the most constant, most frequent, and often the sweetest of our summer songsters. All day, from the young May morn to the full yellow harvest season, you may hear the yellow bunting in the flowery, leafy hedgerows; the reed-bunting on the swampy marshes, and the land-bunting on the grassy marshes reclaimed from the fenland. Their songs differ in quality, yet it is a family song, a short, sweet, pacifying song, a song fit for sultry days, recalling the tone of some æolian harp. From morn to even at this season you may at almost any hour hear some of the buntings, though the reed-bunting sings in the early watches of the night.

But we must confine ourselves to the corn-bunting, or the land-bunting, as he is called down there in the grass marshes, in the roar of the North Sea. There you may see him in the winter feeding in flocks with his congeners the finches and brother buntings; but the oat-stubble is dearest to him; there he loves to wander beside the sandhills feeding mid the cries of the gulls. He loves, too, as do the mavises and partridges, the wet balks of the turnip-fields. But he is scarce worth eating, being rather bitter. And later, when the buds begin to burst on the sallows, you will hear him singing from some bare spray of bramble, and see him flying with his legs hanging down if you startle him, and mayhap you may hear his peculiar shrieking startled note of fright. But he is less common in spring and summer than in winter; perchance he seeks the grassy uplands to build. But still a few remain on the marshes, and when the marshland is yellow with crowsfoot they begin to build their nests in a hollow scraped out in a "bottom," just as does the lark. Indeed the top of his nest is often level with the ground, though at times it is a foot above ground, and sometimes lodged in a hayrick. A coarse nest of dried grass and quicks pulled up by the roots, and sometimes horse-hair, lined with fine grass pulled off above the roots, upon which the four scribbled eggs are laid—eggs that vary much in form and colour and mysterious markings.

All day you may hear the cock crying on some spray, whilst the hen hugs her eggs into life, when the young brood is fed upon butterflies, which you may see the old cock catching as the marshmen mow down the ripe swathes—and a poor food it seems; for, like all buntings, the young are weak on the wing till they are good large birds, even for a time after they leave the nest, and if flushed in a grass field they will drop into the long ripe grass, and are easily caught by the hand. And when the young are full-grown they re-form into flocks and wander over the oat stubbles freshly cut, roosting in the sallows at night until the autumn brings low skies grey. At such season, if you look into their favourite sallows, perchance you may see a flock on the decorative sprays, looming large and soft against the grey mists, a subject worthy of Hokusai. But not long are these pictures to be seen by the cold grey river, for when the hail squalls pelt down the marsh grasses the buntings go into the warm reed-beds to roost along with the few surviving reed-pheasants, sallying out in the dull morning to the corn-stacks, where they get a precarious living; for the gunner is sure to be on their track, sometimes killing sixteen "at a shoot," feeding his ferrets upon their plump bodies or sending them to market by the carrier, the last migration of many a rare bird.

CHAPTER XLIV

THE GOOL-FINCH OR GOOLER.

THIS sweet song-bird is a lover of the ripe yellow wheat upon which the blue lies like bloom upon a plum. And lover of corn though he be, he is dainty, and separates the wheat from the chaff; indeed, his sharp-pointed little bill seems adapted to this end. He is no glutton like the sparrow, who bolts his grain.

You may hear the cock-bird singing his sweet song on a still fine day in the middle of April; and if you stop and listen, perchance you will see his yellow body shining between the new glossy hawthorn leaves; though he is sure to be sitting on a spray clear of leaves, a beautiful yellow spot with dark pencillings against the azure. And when you hear the sweet bunting song, recalling the music of the land-bunting, you will sink down in the hedge and listen with parted lips, for it is music of contentment, filling you with quiet charm.

And a little later, when the irises begin to decorate the dikes with the curved scimitar-like leaves, you may see a pair of birds dragging yellow wheat-straws, sere rushes, dead grass, and horse-hairs to some clump of bramble on a grassy dike-side, and if you search you will see the carelessly-fashioned nest, betrayed by a fringe of straw; and if perchance you walk that way again, you will see the four or five mottled eggs lying cosily on the horse-hair, and you may know that since that bird has begun so early to raise her brood, she will have three or four families that season if all goes well. And should the eggs be robbed by the village urchin, five or six nests will be built, but the eggs dwindle in number, and in the yellow harvest-time you may find nests with but one, two, or at most three youngsters, when the old cock will be singing his sweet song in the dusty hedge-rows whilst they sleep through the yellow harvest moon.

But June and July are their favourite months, and only the more uxorious start their nurseries in April. And the hen-birds are knowing. Do you but disturb one when sitting on her beloved eggs, and watch her fly down the road feigning lameness, or tumbling about with half-spread wings as one dazed with fear. And if, deluded, you approach her, lo! she rises over the hedge and is gone. But far different is it with her young brood, who are weak on the wing for long after they have left the straw-house. Indeed, they can be caught by hand if flushed from a hedgerow and chased so that they scatter over a field. They seem unable easily to rise from the ground.

Young nestlings are at times to be seen in September, but when the waves of cold pass over the many-coloured marshland, the yellow buntings collect in flocks with other buntings and finches and go a hunting for corn and seeds over the fields; and as the clods get powdered with white snow, the goolers draw up to the farm-houses or yard before some barn where the old-fashioned flail is still wielded, the old thresher keeping an old muzzle-loader beside him, with which he occasionally takes a "shoot" at the little flock, selling the chance "rarity" to the guilty "receiver," the bird-stuffer; and so the gooler too mayhap takes his place in a glass case in some commonplace room.

YELLOW-HAMMERS.

CHAPTER XLV

THE REED-BUNTING

THE blackcap, as he is called in the broads, is a marsh bird, loving marshes near running water, more especially in summer-time, for in winter but few blackcaps remain by the reeds and water, but draw up on the land and to the ricks, where they steal grain, in company with sparrows, green-finches, yellow-hammers, and chaffinches; and for joining the company of pilferers, these birds often fall victims to the bird-catcher's net. His appetite in winter is voracious, and as he is tender, soon dying in hard seasons, there must be a great mortality in Arctic weather, when the broads and dikes are laid, and the marshes white fields of snow. Those that remain then—for many go away in winter—often fall victims to the bird-catcher or climate. But in the spring, when the insects come into life, they return to the marshes, and the cocks can be seen popping in and out of reed-beds and clumps of sallow and bramble, their black heads telling strongly against the azure; but at all seasons they keep in pairs or wander singly, never flocking, though they may join other flocks of birds in the winter season.

Towards the end of April, when the waterways are ablaze with kingcups, they choose their nesting-places, preferring a wet-marsh or rond, where either grasses, rush, chate, or water-grasses grow luxuriantly, places dear to the rail and reed-pheasant, though these birds do build in the drier inside marshes and on the walls or in broken-down reed stuff.

At this season you may see the gay cock and sober-coloured hen both building the nest of rotted soft rushes, chate, cotton grass or watergrass and roots, the cradle often being lined with horse-hair. If they build in reeds, they use reed-feather. The nest is merely lodged in the rush, green sedge, or reed, sometimes merely placed on the marsh, sometimes, and oftenest, placed a foot from the marsh, and at other times even two feet from the marsh bottom. When the nest is finished, the hen lays her three or four, rarely five, scribbled eggs, and forthwith begins to sit, the handsome cock relieving her at intervals. When in the fulness of time the black youngsters with their gaping red maws are hatched, both parents are busy, from dawn till late at night, gathering food for the gaping maws, never alighting on the nest, but dropping into the stuff near the nest, and creeping to it under cover, feeding the young in the cool shade, for the opening of the nest is turned away from the sun, if possible, to protect the young from its rays. Indeed, this is the practice of most birds. The bird is tame, or "cottly," as they say in Norfolk, yet full of deception ; for if you approach the nest, the sitting bird will run or fly off as if wounded, get her back and feathers up, flap and "trapse" (trail) her wing, trusting you to follow her, and leave the nest. Should you be close upon her nest, she will keep rising and hustling, leading you away from the precious nursery, even falling to the ground with outspread wings, helpless—so artful is she.

The young are fed by both birds on insects gathered by the water-side, especially from water-flags, and on flies and insects picked up on the marsh bottom.

They raise two broods a year, if not robbed ; but should anything happen to either clutch, they will go on laying, so that eggs may be found after harvest even, for there are then plenty of maggots, cankers, and midges for old and young.

When not engaged in rearing their young, they sit upon "sprags" or reed-stalks, after the manner of the land-bunting, and sing a somewhat harsh song, like that of their blood-relatives.   All day long through the sultry summer from day-break (for they are the earliest birds to awake in summer), their simple, plaintive song may be heard—even through the hottest noontide.

They are simple-minded, monotonous little citizens, rather commonplace in appearance and character, but unassuming and harmless; and there are many birds we could afford to part with more readily.

REED-BUNTING'S NEST AND EGGS (taken *in situ*).

CHAPTER XLVI

THE STARLING

THE starling is a dirty bird—dusky-skinned, gaily-spotted like a dung-fly, fruit-thieving and imitative. A hanger-on to the borders of civilisation, he has learnt all the petty meannesses of the Broadsman and none of his noble qualities; he is a filthy pariah, a lover of warm chimney-corners and animal droppings, and his song, now thrush-like, now recalling some finch, is stolen; he is a born plagiarist, a dirty, sordid little creature, and full of the citizen's cunning. I never knew him do a kind deed nor an unselfish one; he will not even fight for his young; but, like many contemptible little things, he, in certain lights, looks beautiful when he is arching in the liquid azure in autumn flocks over a reed-bed, or when he is adorning the autumn reeds, weighing down the golden stems to the still water's bosom.

In early spring, when the marshlands are dry and bare, you see him flocking with rooks and jackdaws round the cattle and sheep just turned out to marsh, alighting on their backs, feeding on their droppings, flying greedily along, the tail of the flock flying ahead of the leaders, so advancing with gobbles and twitters in alternate battalions over the marshland and soft holls, prodding for worms.

And if the weather be open, you will see some solitary bird singing his borrowed thrush-like song from some tree-top or mill-head, flapping his wings at times, for they pair according to the seasons, some earlier, some later; indeed, I have seen unpaired flocks ranging the marshes till May-day.

I

When paired, they select some hole in a tree, some crevice under the fenman's cottage, where their four or five commonplace eggs are laid, and then both birds go busily a-marsh for wire-worms, earth-worms, grubs, moths, midgets, cankers, and butterflies, to feed their greedy, noisy, rude nestlings, their noisy chatter and din crazing many a cottager so that he will often shoot one or both. Should the hen be shot, the cock still goes on bringing up the garrulous young, for they do not leave their nest before they can fly well enough; they are too careful of their skins for that!

As soon as they can fly, away they go, old and young, in noisy flocks, to the marshes, to live on cattle excrement and parasites, or occasionally to make incursions into the fenman's garden to steal cherries; but the old marshman is soon after them with a curse and an ounce of shot, killing or maiming them. But then a cat won't touch them; they are such vermin. Pussie will scarcely deign to kill starlings for fun, and he certainly would as soon think of eating them as you would of eating carrion crow, although, on the other hand, I have known old smelters make pie of them.

In the early summer you may find them up at all hours of the night, for they sleep upon the grass marshes near the flocks and herds till the young reeds are tasselled just after harvest, when they take flight to the reed-bed, and 'tis then the damage is done, for their filthy bodies weigh down the dainty and elegant reed stems, and break them. That is, if the watchman of the marsh crop do not, with rusty old muzzle-loader, keep firing at dusk to scare the circling flocks away. One starling settling on a young harvest reed will break it in the middle and so cross-lay it; but when the reed is sere and ripe, three or four birds will not injure a stalk; and the watchman knows this well, for he only guards the swamp crop whilst young. Later on, when the sere reed is circled with ice, he does not disturb them, and you may any winter night, just as the red sun sinks adown the grey sky, see flocks upon flocks come spreading across the grey for

perhaps a mile, all flying from the cold marshlands or fenman's stacks in parties which circle around over the reed-
bed a few times and then sink into it with innumerable
chatterings to sleep through the icy night-watches whilst
the water freezes hard and clear below them, and the tassels
turn to frosted silver above them; for the starling is one of
the hardiest of birds. But even he is of use to the fenman;
he is a lover of wire-worms, and will follow the plough
steadily day by day in search of the hideous, foul, tough-
skinned wire-worm, that lover of carrots. And though,
like rooks and wild-fowl, he himself is alive with lice in
summer, he is sedulous to pick ticks from the flocks and
herds, thus again befriending the farmer—scavenger that he
is. In August, too, you may see him hawking for flies after
the manner of a swallow, clearing the air of vermin as well
as the land.

Low down in the dazzling chalk coast of Kent the starling nests with the house-martins and jackdaws and rock-
pigeons among the bright yellow wall-flowers and rosy vale-
rians. And that is why, perhaps, old fenmen in Norfolk
will tell you there are two kinds of starlings—the wood-
starling, which nests in houses, and nests earlier than the
rock-starling, which nests in cliffs; and they say the
wood-starling is the one to talk, "if you teach him," for
the starling will become the tamest and most docile of
birds if taken young. One old man I knew kept one for
years, not even clipping his wings. He named him Billy,
and Billy, like a good boy, always answered to his name,
going up to his master.

Billy's master was an old ratter, and his ferrets would
eat starling; so often on a winter night old Bob would
go down to the reed-beds, where the birds roosted, several
huddled together on a reed for warmth, and shout, and
the mighty flock would fly out of the reeds with a roar
and thousand-tongued chattering, when the old muzzle-
loader would flash red and roar in turn, the shot cutting a

"sheer hole trew them so you could see the sky," as the old fenman and ratter said gravely. He knew, too, the fir-trees where they slept after the reed-beds had been cut.

And so the dead scavenger goes to feed the ferret, and the ferret goes to kill the living scavenger—the rat; and thus the great fly-wheel of life revolves.

CHAPTER XLVII

THE JAY

WHEN the coppices resound with the challenges of the cock-pheasants, the surviving jays will be heard shrieking their sharp, shrill notes, that startle every head of game, sending the challenging cock-pheasants into the under-growth, and startling the hares from their forms, often driving them straight towards the intruding wight. Far and near, over the still waters of the lagoon, you may hear that shrill voice in early spring when the gladen is green, in summer when the lagoon is pied with lilies, in autumn when the lily leaves flash like bronze cups upon the waters, and again through the icy mists and snow squalls of winter, when the gull seeks the peacefulness of these inland waters.

And should you neglect to preserve your coppice by the riverside, no challenge of cock-pheasant will be heard, for the blue-winged jay sucks every egg on principle, even to the six blackbird-like eggs of his own tribe, which are laid in a nest like a very deep wood-pigeon's, only the jay lines his with yellowish roots. If a small planting hold but two pairs of bright jays, there will be such feuds in the spring when the hawthorn 'gins to bud—such shriekings and com-motion, such a searching for one another's clumsy nests and eggs, and such shrill shrieks of triumph when the eggs are being sucked, that make the leverets tremble in their lairs, and the wood-pigeons to fly hastily to and fro. And know-ing their love of egging, the wise keeper baits his traps with eggs, and so captures the bright-plumaged dare-devil-looking birds, unless they smell the traps, and flee with that heavy slow-beating flight of theirs to a neighbouring and keeper-less planting, that game knows not and where all things are amiss.

And when the covert is secure to himself and a bold missel-thrush mayhap, and his young are hatched, you may see this terror of the wood hunting along the hedge-rows for helpless, yawning nestlings or feeble flappers, for all is grist that comes to his mill at that season, especially redpolls and linnets; and the flappers know it, for they hie to the hedgerows for cover, like startled mice, as soon as the shrilly-voiced ranger darkens the sky, which is fre-quently, for he eats little and often.

And when the fledglings are full flappers, the old fenmen sometimes tame them, and by slitting their tongues give them the gift of speech; and apt pupils they prove, some imitating the cackling of his mongrel hens, the barking of his still more mongrel curs, and even his own harsh voice, saying, " Jacob, poor boy," or calling loudly in unknown accents, " Boat ahoy ! " imitating the foot-passengers' voices at the ferry, and answering himself as doth the ferryman— " Ho ! ho ! "   Nearly all things he will imitate but the cuckoo—that roystering bird is safe from being plagiarised. Ask the titlarks why ?   And when the oak-trees hang heavy with acorns the jays are happy, for they love acorns, holding them with one claw and eating them as a monkey does a nut.   So precious are they that they will store them under the earth, stamping their treasures down as carefully as any pirate might his plunder.

But the keeper is ever on the jay's track, and he is grow-ing scarce; and so the world grows duller day by day as the gay bits of colour are being killed in order that game may flourish.

CHAPTER XLVIII

THE JACKDAW

THE crafty "cadder," as the Broadsmen call the jackdaw, is by no means common in late spring and summer, though in early spring and autumn he is frequently to be seen on the marshes in company with starlings and rooks, all paying delicate attentions to the herds and flocks, perching on their backs and picking off the lice they harbour. In this wise the cadder is the farmer's friend; but he makes the farmer pay for his slender services; for he, too, is a lover of young corn, as also of worms, young birds, pieces of fresh meat, and the farmer's strawberries. In winter the cadders are black on the land, for many have come over the seas with the rooks and starlings; but in spring few are left, and these, crafty as any priest, stick to the church, building their slovenly nests in the steeple, where the young blue-eyed cadders are reared.

But the cadder is fonder of the cliff-land. The fenmen are too sharp for him; he fares better in the cliff-land, where the men are of his intelligence, and as fond of the " silver " as he is of the strawberry patches near the Kentish chalk cliffs, where he often breeds.

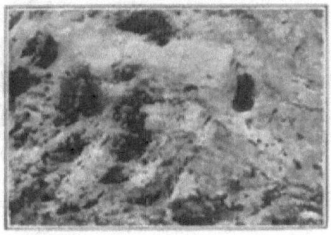

JACKDAW'S NEST (DEVONSHIRE).

CHAPTER XLIX

CROWS—BLACK AND GREY

NEXT we come to the black crow, a distinct bird from the grey crow in my opinion, with all due deference to what has been said to the contrary, though, as is well known, the two breed together.

The black crow never flocks (although a clutch may be seen together in autumn); the grey crow does flock, sometimes in large numbers.

The black crow is resident in Norfolk; the grey crow is a migrant, though isolated specimens do stay throughout the breeding season, but I never heard of them nesting.

The black crow loves the woods and plantings; the grey crow loves the open marshes and mudflats.

The black crow is quicker and shyer than the grey crow, and more restless. The grey crow will sit for hours in damp weather, moping like a wood-pigeon. Although inaccurate literary hodmen have said they never keep still, any gunner in Norfolk could teach those pseudo-scientists better.

The black crow is rare in the Broadland, for coverts are rare and gunners plentiful, but occasionally he is to be seen; and one pair I knew of built in an old rook's nest in a rookery in a small planting. Immediately you entered this rookery, you saw the pair fly up quickly and flap their wings, hanging over the trees for a time, then flying away and returning again, but flying off and coming back no more until you had gone, so shy and artful were they.

You may know the black crow by his blue eye, like a jackdaw's, for his feathered bill is no test, the rook having at times brown feathers; and you may know his egg from a rook's, for it is larger and bluer; and you may know him amongst rooks by his flight, which is quicker, and his actions are less sluggish, nor does he wheel about so much as a rook; and when seen near his children, he soon leaves until the strangers go away.

But he is rare in the Broadland, as I have said, and nearly a thing of the past.

THE GREY CROW.

Not so with the "Kentishman," or grey crow, who is one of the commonest winter birds of the marshland—a brave freebooter, in handsome uniform, who ranges from the sea to the upland, through snow-blast and sunshine, a spirit of the lone marshland, one of its winter voices.

Indeed, he is one of the most familiar birds of the district —a handsome, rather sluggish bird, dressed in black and grey, with a powerful black bill.

Soon after Michaelmas, when the Broadland is a harmony of blue and gold, the grey crows arrive in large flocks, and, separating into small parties of fifteen or twenty, they scatter over the Broadland, some of these flocks dividing into smaller parties, finally dividing into pairs, or even going singly; and when nearly all the bird life of the district is starving with hunger, the grey crow is busiest. All winter long, whether in sun or hail, in sunshine or fog, you will of a morning see the grey crow leave the low alders by the broads, where they roost, near to the rooks, and go with sluggish flight to their beats, for they are late risers; nor are they so active as a rook, moping as they do, on grey foggy days, like wood-pigeons—an altogether different bird from the active, alert black crow, who is never still save when he is dead. And if you follow these birds in the morning to their hunting-grounds, you will see them range along the land, you will see them about the marshes, round the bottoms of stacks, on river walls and heaps of stinking keel-muck, sitting solemnly round the open waters on the frozen ice, or by the edges of running rills, near the snipe, when all else is bound in icy fetters, all on the alert for food, for living prey or carrion. Let a wounded duck fall on the clear hard ice, and the old Kentishman will fly up and seize it before the gunner can break a way to his prey, and, turning it over, the bird will bury his powerful bill in its living, bleeding breast, just above the breast-bone, and tear open its quivering wind-pipe; and if the angry gunner is not revengeful, he will pick the duck clean as a party of ants, even to his eyes, leaving only his head. No bird can pick a raw bone cleaner than a Kentishman. Or if there be no ice, and the stricken bird is not dead, he will seize him, and carrying him to the shallows, amongst the formless dead gladen stalks, he will "strip him out," as the Broadsmen say. Rats in the stacks, mice on the marshes, all have felt the force of that cruel bill, as they were pulled to bits; and rabbits on the sandhills, hares drowned in the icy dikes or left dead in snares, the old saddle-back crow cleans them all up, and polishes his shining bill complacently afterwards upon some bare branch by the water. And horses drowned in the icy ditches, he and the pick-cheeses will clean them up and leave their bones mouldering on the marshland; dead and swollen fish turned up and killed by the "salts," and left putrefying on the ronds with filth and drift weed, he and the rats clean them all up. Any carrion he takes; he is not particular, as you may see by watching him at a stinking heap of keel-muck that breathes pestilence by the river-side.

Nor are they shy. They frequent stack bottoms often, hunting for rats and mice, which they will eat dead or alive; and should you throw them bread in hard weather, they will eat it; and when the weather is very hard, you may see them, along the frozen roads, pulling the steaming horse-dung about, extracting food, after which they will beat heavily over the marshes in search of carrion or mice, which they swoop down upon, never hovering like a hawk ; and if food fail them there, they take to the sea-beach or mudflats, feeding upon dead fish and worms ; or else they go to the stacks and rob the farmer of his corn, pulling it out *secundem artem ;* or else they hie to the green turnip-fields for worms and slugs, and in an Arctic winter they will eat raw potatoes and turnips to stay their pinched stomachs. And when the blackthorn is in flower, they frequent the lambing pens, pecking the eyes from sick or sleeping lambs, or tearing the tender flesh from the dead young lambkins that lie by the dike-side. At this season, too, they follow the bottom-fyer, proding the soft black mud for small eels, worms, and fish scooped from the sluggish dikes ; or else they are preying upon young rabbits or leverets, or cutting at the early-laying peewit, whose eggs they love as well as any gourmet.

But with the first of April they begin to move towards the warrens, where the first peewits lay and the young rabbits live, and you may see flocks of them sitting about waiting for a southerly or sou'-westerly wind, when off they go, flying up to a height and starting off to travel over the sea. But all do not go at once. Some stay to enjoy the early eggs and a few dishes of young birds or frogs ; but when May comes in most have gone, but not all. Every summer a few remain, but not to breed, so far as is known. I have seen them by the sandhills, and on Breydon, in every month of the year, sometimes in company with rooks. All winter through you may hear their hoarse cawing—bird calling bird through the dreary landscape. "Quah, quah, quah," a cock will call, and his rival, some hundred yards off, will answer in the same tone of voice, "Quah, quah, quah." Then his voice will grow shriller, and he will call quickly, "Quah, quah, quah," and the rival will again answer in the same tone of voice, till both, tired out, will cease calling. And on grey days you see them sitting on bare trees, silent and grim as the dead.

A useful bird is the handsome but sluggish "saddle-back;" as indeed most carrion-feeders are sluggish, though the superficial have called the grey crow an active bird, who is never still. Methinks they judged the grey crow for the black crow; quite a different bird, to all but a faddist—that is to say, an "outdoor" naturalist.

And would it be believed the saddle-back is eaten? rather secretly, it must be confessed, but eaten he is and in London. Kentishmen are sold and eaten as "sparrow-duck," and I daresay they are good enough, for they are no dirtier feeders, if as dirty, as the domestic duck, whom I have often seen gobble down fæces. Truly we are a strange people; the duck and pig we cherish, whereas the "sparrow-duck" we throw aside as carrion. Truly matters of taste are not to be argued about.

CHAPTER L

THE ROOK

THE burgher amongst birds, he has all the faults and virtues of his class. He has foresight, cunning, and organisation, but he is vulgar, greedy, and commonplace. A thief from little birds, a coward before fighters, he is a true representative of the big commercial citizen; also he is fond of his dinner and greedy enough for an alderman.

His (and her) carriage, too, is that of the plump, well-fed, graceless citizen, ay, even to his bourgeois gait and strut, and lastly, his voice is harsh and uncultured, which completes the simile, unless we add that he is eminently philistinic and respectable, which, poor brute, he cannot help.

In the early spring, when the buds begin to swell and the trees are greenish of hue, the rooks return to their city, the old birds driving the young away to find a colony for themselves, for a rook city does not increase quickly. There you will see them sitting on their nests or repairing others with little dried sticks, or even building new ones, which only take a few days, if the elders permit it; for theirs is a truly conservative government, and should a pair build without permission, the old birds will go and pull the nest to pieces with vulgar cawings—a regular city riot. Nor are they particular where the nests are built; they build in low cars on the marshes, a few feet from the ground, or on the slenderest and most insecure-looking branch near some upland farmstead; they do not, as poor Richard Jefferies asserted, choose *a strong fork;* they are by no means so careful as that, and often build eight or ten nests on a slender branch. They are punctual, too, in their building, as become burghers. No matter how hard the winter or tardy the spring, there will always be young rooks on May 14th. Even after the Arctic winter of 1890–91, there were young rooks on the appointed day.

Much has been written as to the time they pair; my own impression is they pair for life, and, like the citizens they are, they are most discreet in their love-making; it is done so respectably, away on the marshes, as a rule, in a secluded spot recently mown, and always on the ground, and early of a morning. They don't like to be seen, but I have been bold enough (or rash enough, an you will) to surprise them in the act. There is no rapture in it, but it is a mere business-like proceeding, eminently respectable and bourgeois.

As will be seen, they begin their nesting in February, soon after appearing at the rookery, with jackdaws and an occasional artful carrion crow, whence their noisy cawings resound all day.

When the young are hatched, business commences, and the old cock-birds go off to the fields and marshes, gathering pouches full of wire-worms, grubs, beetles (from old cow-dung heaps), returning ever and anon to the rookery, over which the hens are watchmen. The worms they capture as does a thrush, but they are quicker, and do not take two bites at a cherry.

If at this season of the year you approach a rookery, you will see the birds rise above the trees before you get near, cawing and flying about in the air, signalling to every rook in the country round that that bad man is coming; but if you steal up and lie in a dry ditch, as I have done for hours, they will gradually settle down, but you may be sure more than one pair of bright eyes is upon you to signal your slightest suspicious movement.

You will then see the cocks come flying in with full

pouches, that makes their voices so hoarse that you can ever tell by his caw who has brought back a good bagful of provender and who has not; and you can see their distended throats, moreover. As soon as the father of a family enters his tree, his wife flies to meet him—nor does she always wait till he comes into their tree—and takes the food from him. You may see them doing this on the wing, but it is generally done on a stout branch. Then there is such a noise and cawing as the young birds are fed in turn, their mouths wide open, and their necks outstretched, and their wings flapping, as do the wings of the rooster. Turn by turn she feeds the voracious youngsters. When you are tired of watching them, and arise from your cold bed in the ditch, the signalman caws, and away go the elders cawing and calling, flying over the rookery, and shaling round, whilst the youngsters join in the vulgar din, which is kept up until you get far away from the rookery.

It is at this season of the year that the rook does so much harm to the farmer, for though, like most birds, he does good too, his harm far outbalances his goodness, and farmers would be better without rooks, notwithstanding the "bird-lover's" statement to the contrary.

Let us briefly try to balance the matter, and see whether the farmer be on the debtor or creditor side, to say nothing of the game preserver.

Taking him all the year through, his staple foods are, in this district, grubs, worms, wire-worms, beetles (from horse-dung), frogs, eggs, young birds, corn, turnips, mangolds, potatoes, and barley, besides food found by the shore, and bilberries in heathy districts.

When he follows the plough in early spring with the lapwings and black-headed gulls (not grey gulls, as the superficial poet puts it, for they never follow the plough to my knowledge), he undoubtedly does some good in eating grubs and wire-worms. His eating earth-worms does no good to the farmer, neither does his consumption of

harmless frogs, eggs, or small birds, for he does not eat
young sparrows or greenfinches as a rule, but the more
harmless rails, waterhens, and often game eggs and birds;
but we are coming to that. But he does consume wire-
worms, you will say. Yes, and on the marshes he does
good so; but what is it when you come to a lot of turnips,
or young corn, or peas? If he do not steal the grain in
the ground, which the clapper-boys calling, "Car-whoo!
car-whoo!" through the grey mists endeavour to prevent,
he will, in his zealous search for wire-worms, go into a turnip
or mangold field and pluck up the roots ruthlessly to see if
there are any grubs or wire-worms at the roots, for they
don't eat the roots. A flock will soon spoil £10 or £20
worth of roots so—ay, in an hour, if left undisturbed. Sometimes the whole crop is so spoiled, and the farmer has to
set the field again. Or again, watch them in a cornfield,
if you can elude the watchman who sits upon a tree-top
over the field signalling to the others below.

Then you will see them pecking the ears of corn, which
they carry off to a wall to eat in security. But you must be
sharp for the work, for the stealthiest gunners have great
difficulty in shooting them when at that game. An experienced hand told me he once crept down a furrow on
a marsh cornfield with his gun; directly he got within a
hundred yards of his prey, an old cock flew over him and
began calling "caw-caw," and away they all flew cawing
triumphantly. The following morning, at daybreak, he
repeated the experiment, and though they were less
cautious, he only shot seven as they arose. In the young
corn they are more mischievous, and will soon spoil a
couple of acres of blades. They go over the new-lays and
pull up every weakly plant, and any plant turning yellow,
to seek for the worm at the roots. It may be for this
reason they pull up healthy roots growing amongst the
withered hoed roots, for they see the withered thinnings,
and may think the others have worms and that the dead

roots have been killed by the worm instead of the hoe. A
thousand crows in a barley-field is no uncommon sight.
But if you are passing you may scare them away, but not
by shooting them and hanging them on a stick, as is often
done; for many is the time you may see them feeding
quietly around such dead mates. No; you must shoot
your "crows," as they are called in Norfolk, and lay them
flat on a hillock in the field with outspread wings: that
will frighten others away—they think they will be trapped;
and when one realises the vast numbers that flock over in
autumn, any method of scaring them is invaluable, for
though some of the home-bred rooks go away across the
seas in autumn, far greater numbers flock in to steal
English corn, preferring that planted on the marshlands,
perhaps because the rook is fond of water; indeed in great
drought the rooks die like flies—they cannot live without
plenty of water. To game they are very destructive, sucking
the eggs and eating the young.

After the rook shootings, in which the young are usually
thinned, in July, the whole city of young and old desert the
rookery and take to the marshes, and then the young birds
are nearly as large as their parents. It is long before they
feed themselves, however; for there on the marshes you may
see the old birds feeding the young with their boluses of food
after the manner of a pigeon, but the rook is quicker in the
operation than the pigeon.

When they have left the rookery, some return there nightly
to sleep, but the majority choose another roosting-place—
some place that will be warm and cosy in winter—often a
low car by the river or lagoon side, roosting in the low
trees with Kentishmen, though they differ from their neighbour in not eating carrion.

But the rook is a coward at heart. He behaves well when
stealing, and his watchman sits aloft, turning and spreading
his tail feathers every time he passes the caw "All's well;"
but should a sudden shrill sound break through the still air,

K

there is a stampede ; bass sounds do not seem to frighten him. Indeed, he does not seem able to hear a bass noise like a shrill treble. If you listen to them, you will hear the cock-bird calling "Ah! ah!"—a signal of arrival at the city gates; then the hen calls "Ah! ah!" as she takes the supplies and flies towards the young; when they begin a frog-like noise, or crying like babies at the breast, as she feeds them. On the other hand, the danger signal is either a peculiar gull-like noise, " *Kёo, kёo,*" or a " *Quah, quah.*"

His life is hedged round with precautions, the precautions of a mediæval city; but once he gets ranging over the marshes, his true character shows itself; for a lapwing, whose eggs he loves, a redleg, a kestrel, a sparrow-hawk can all easily chase him away. He flies and never shows fight. His cunning, too, has taught him how far a gun will carry, and if he cares to rob your garden of walnuts (which he dearly loves) it is at daybreak in the autumn; if it be your potato patch, he turns up when there is no watchman present, and he seems to know when the allotment gunner is present to protect the crops. He is artful, and he knows it, but withal he is a burgher even to his mourning when a mate is shot; for he will then cry distressfully, and shriek and howl over his dead mate, lamenting with vulgar noise the civilised thief who has gone down. When the eagle was made king of birds, the rook was elected Lord Mayor.

A ROOKERY IN SPRING.

CHAPTER LI

THE SKYLARK

In April, when soft showers draw a fine gauze over sky and marshland, the larks are most joyous. At this teeming season, when a silver shower has washed the bright warm air clean, the larks will go hovering until they look no bigger than a fly on the face of the pale silvery moon. You see them rise in numbers from the level grasses with quickly beating wings, their tails opening and shutting up like a Spanish lady's fan, beating the soft zephyr as they sing triumphantly in their upright course to the stars, passing up a mill-height in a moment, carrolling merrily as they mount through the still air in their vernal flight, which is not so swift as a titlark, yet more graceful; and if the day be fine, they are soon lost in the azure; but if there is going to be wet, they do not go higher than the mast of some tall ammiral, but hang in the air, hovering like a kestrel, foretelling a storm to the country folk; but after a spring shower, when a fresh breeze is blowing that has gladdened the flowers, they delight most to mount or soar in ever-widening circles as pleases their fancy; then, say the fenmen, it is going to be fine.

And if you lie on the damp water-grasses on your back and gaze at the little speck of life as he descends, you will see him sink forty yards by forty yards, and at the end of a strongly marked musical flight you see him, when within a mill's height of the ground, shut his wings and tail and drop like a stone towards the green marsh with a trill, opening his wings and beating the air a foot from the ground, soaring a foot again ere he drops and squats in the grass, as is his custom on meeting a stranger.

If ever he sees you walking over the soft marshes, like the partridge, whom he resembles in many things, he will squat and trust to capricious fortune. A breezy day, too, is dear to him, for soaring is the easier; he rides upon the wind in glorious freedom. But if you are not to be seen, woe to the brother lark who catches his sight when he alights; for his native pugnacity is aroused, and he darts at his foe all quivering with excitement, running at him eager for the fray; for larks seldom walk, except upon a "new-lay" and in cold weather. In fighting they will often fly up face to face, fighting with beak and claws and beating each other with their wings.

At times, too, out there on the silent empty marshes, you may witness a tragedy as a young lover soars into the azure and suddenly drops dead from the heights beside his mate, rebounding from the marshland like a stone, his little heart being broken and his song having silenced in mid-air.

All the year round the lark is the usherer in of the day. No sooner does the daylight sky rise than the little balls in their grassy forms start into life, for at eventide they puff into a ball ere they go to sleep. Dung is dropped, and they begin to sing, some soaring into the starry sky, others answering with a defiant note from their grassy beds. And so all through the day you may hear them at the proper season till dark.

And in June you will know that the courting season is near, for you will then see trios over the amber reeds, two cocks and a hen. Neck and neck they fly, one cock chasing the other, and darting every now and then at his rival, but seldom pecking him, and the flouting hen enjoys it, flying as strongly as they, uttering a peculiar note. Round and round, over reed-bed and dyke and marshland, until one cock flies off and leaves the victor with the hen. They, later on, fly off to a grass marsh to collect quick and foul

grass to line the cup-shaped hole they have dug in a tuft of
grass in a dry marsh, and afterwards they will line their
nest with dried grass, and she will lay two, three, or four
reddish eggs as her fancy dictates. Later, if you go to the
upland, look for the lark in a turnip-field, or in the clods
of a wheat-field ploughed fallow.

On a fine day during the pleasant season of love-making,
when they are not soaring, you may see them taking a dust-
bath, just like an old hen or partridge; or, at another time,
after a shower, you may see them bathing in pools on the
marshes, happy as larks.

And when the serious duty of sitting begins, he takes up
a station near the nest, which he watches closely as any
lover, or you will see him sitting sentinel on a decaying
marsh gate-post, or upon a grey-green sallow stole. And
should you draw near, you will see the old cock hover
about you and sing with affected indifference, while the
hen flies off like a peewit. And should the eggs escape
the thieving mouse-hunter or weasel, you may easily find
their nestlings, for the cocks hover with food over their
nests like a titlark, and betray their nestlings to the practised eye.

And when the young, who resemble young partridges
more than any bird that flies, have arisen to the dignity of
" full-fledgers," they gather into flocks, frequenting the new-
lays and turnip-fields, and then the mischief begins, and
continues all the year until the spring comes round again,
for they all love the uplands as much as the marshlands.

And when the partridges are ready to shoot, they will
lead many a dog a fool's chase, for they have the " scent,"
ay, and the very flight of a partridge. At that season, too,
when the rime crystals deck the ebon chervil stalks, thousands from across the seas increase the marauding bands.
You may see them land—the wheat-pickers, accursed of
the farmer—when the wind blows shrewdly from the east,
or the more temperate south-east, for they, like crows and

other migrants, choose a beam-wind for travelling. And herring-fishers will tell of the brave flocks flying slowly across the grey sea, the wearied alighting on the treacherous waters, thinking it a new kind of marsh, perhaps, only to be engulfed, or others, stronger, will fly up dripping brine and spreading their sails, go on till they topple giddily on to the wet sea-beaches.

And when the reeds are yellow and the waters cobalt, the farmer's heart is aggrieved, for then bands of hungry thieves feed upon his newly-planted corn, pulling up the plant and eating the succulent roots; or else they steal the hearts from the green clover plants in frosty weather; or, when the land is white with snow, they will pull off the young turnip leaves just escaped the fly; and then thirty fall at a shot of the old muzzle-loader, for at that season they are easy to shoot. When on the wing they are more difficult than a snipe to bring down, even to the experienced.

Alas! that it should be written, but the lark is the farmer's greatest enemy, not even the burly rook excepted. And though hundreds die in hard weather, even in the threshed ricks at the farmer's doors, and many perish a prey to stoats and rats, who love them as dearly as any gourmet, dragging them to their holes to eat them at leisure, leaving the wings as a tally; still there are more than enough to provide stock for future families, and the surplus are far better killed to be served on toast, which is a dish too good for any born sentimentalist, and fit only for a sensible naturalist. In the azure carolling or on toast beneath the pale yellow rays of the winter lamp they are equally delightful.

CHAPTER LII

THE SWIFT

THE largest of the swallow family is the swift, or " Develin," as he is locally called—a big, black, mysterious bird, that arrives at the end of May, last of all the swallows, and is to be seen flying high or low, according to the weather, but always to windward, always to windward, foretelling bad or hefty weather or rain, according to local tradition; for the swift is rarely to be seen in the Broadlands when the weather is fine and settled; indeed, they say *là bas* that they all go away somewhere. Whether that be so I know not, but when you do see the develin there, he is pretty active, especially when fighting in spring, for they are good fighters, having many a good round over a poor moth's body, shrieking their shrill cries all the time, flying up and down the welkin, but rarely in very large parties. Sometimes one bird will be seen beating to windward, sometimes twenty, but once I heard a large flock right up in the sky above me, almost in the clouds. Can it be that in fine still weather they go up so high that we do not see them?

The swift is a mysterious bird in the Broadlands, who is said to nest in steeples, but his nest I have never yet found, and the little I know is that he is there one day and absent another. But they are always plentiful during hot summers. Indeed, all the swallow family fill us with the sense that they are birds of passage; they always seem on the move—here one day, there another—for ever travelling and eating and drinking; a mere phantasmagoria of bright and restless atoms, for ever flickering athwart the skies. Such is the impression of this strange fly-eating family, for whose good visits one cannot be thankful enough, and whose lives a very heavy penalty should protect, for they are the most useful of birds.

I have known them stay in Anglesea till the middle of October.

CHAPTER LIII

THE NIGHT-JAR

THE night-hawk, or big razor-grinder, as he is more rarely called in the Broadlands, is by no means common, for cars abutting on the water are rare, and the night-hawks love best the solitude that dwells in a car by day, for he does not care to be disturbed as he roosts with his speckled body parallel with the tree branch, or rests upon the ground—that is, when he is roosting—for sometimes of a dull day he will hawk by day.

But the dusk is his chosen time; the hour when the big bats hawk over the water sees him fly beneath the moon from a sleeping covert, just as the partridge is calling her young on the uplands near by, the heron flighting to the dikes, and the last snipe drumming round and round over the moist water-grasses.

Any fine day in May you may first hear his "razor-grinder" going, as the Broadsmen call that strange jarring voice, as he hawks over the reed-beds for moths, combing their hair before he eats them with that long serrated claw of his, as the fenmen say.

Later in June, when they have grown more common, you may flush him from the marshes, whence he rises awkwardly, and flies heavily off to a tree, or drops again into the stuff far ahead of you. An old fenman followed one, one hot August day, to a thorn-tree growing on the marsh edge, and he found "him sitting all along the branch, as if he was glued to it."

They are fond of the same locality, and will return year after year to the same neighbourhood, though their eggs and young are seldom found.

But this strong, slim, simple bird, with the colours of the night—lilac with the setting sun diffused through it—is seen at his best as he hawks—like a dumb machine—at closing-in time, or at dawn over the reed-beds, flying low over the gladen or reeds, darting quickly at the yellow underwing moths he feeds upon, or hawking up and down, filling his crop with midges, after the manner of black-headed gulls or terns; indeed, in the dusk it is difficult to tell which of them is hawking in the pure air above the reeds; but directly you startle him, and he gives voice to his peculiar startled cry, you recognise him to be the loud-mouthed night-hawk; nor will any artifice lure him to you, for he is wiser than the owl, of spurious reputation. And after his meal he goes to roost in the gladen, and later, when the flight shooters turn out for full flappers, the night-hawk will be their companion, hawking close to their guns for moths, until the winter equinox blows them across the sea, leaving a strong young bird here and there late into November.

A lover of chimneys, too, is the night-hawk, and when walking along the coast you may see them flying round quaint old chimney-stacks in a gale, awaiting a fair wind to waft them across the grey sea, where he will resume his churning whilst hawking as well as whilst sitting on the ground—for he gives voice during both acts—using his "big wheel" and his "little wheel," as the Broadsmen distinguish the two kinds of churning for which he is noted, noises to be heard all through the short summer nights. A mysterious bird of night, bearing the sombre colours of the reed and the night upon his body, and bearing in his record the legend of goat-sucker, the etymology of which I think is at fault, unless, indeed, the *goat-hawk moth* was meant, and the bird originally called "goat-hawk sucker," and subsequently "goat-sucker;" but the Broadsmen know nothing of this widely-spread superstition. Was it like the name of "reeler" for the grasshopper-warbler, invented by purblind men of books who wander in some *hortus siccus?*

CHAPTER LIV

WOODPECKERS

THE green woodpecker is a lover of water; indeed, I have never yet seen him far away from the water, whether it be river, lake, or the sea.

There stands an old alder by one of the silver sheets of broad water where the greenpecker has raised her young more than once, and where on a grey day in autumn you may watch the green bird twitter up the two paths round about the trunk, flying off with a loud pheasant-like cry, when disturbed, to a distant cover.

In winter I have frequently disturbed the green woodpecker feeding by the low cliffs by the sea, or upon the sea-sand itself, in search of insects, no doubt.

In Anglesea I have seen it feeding on a sea-flooded lowland in company with curlew and green plover, and not far away were oyster-catchers and peewits busy amongst the oysters and worms. When flushed, he invariably flew off to the nearest hedgerow or cover.

But I have had mere glimpses of this bright, loud-voiced bird, for he is very shy, though not very alert.

LESSER SPOTTED WOODPECKER.

A long lane, bordered with pollarded willows, leads to the desolate hollows of the sandhills, beyond which is heard the everlasting cry of the winter sea; for this lane is of interest to us only in winter-time, for 'tis then you make sure of seeing a pair or two of the lesser spotted woodpeckers feeding and climbing restlessly about the rotted bark of these willows, now shorn of leaves—a living gem against the silvery background.

Year after year, after the autumn equinox, you may see these birds thereabouts, but after the March equinox they are gone; they do not stay to breed, but seem to migrate elsewhere to rear their young.

And this is all I have seen of the lesser spotted woodpecker in the Broadlands.

THE GREATER SPOTTED WOODPECKER.

The greater spotted woodpecker I have never seen in the district, but an old keeper told me he has shot them, and they used to frequent a planting near the water. He said he had a whistle similar to the green woodpecker, and when he pecked at a dead branch, he made a peculiar vibrating noise when pecking. He used to decoy them into the open by placing sheaves of reed on the top of a bramble bush, and when they came to them he would shoot them; he avers he has shot several by this plan.

CHAPTER LV

THE KINGFISHER

QUAINTEST of shape and gayest of colour is the bright kingfisher, who, with his dreamy eye, may be seen perched with grotesque gravity upon some willow branch overhanging the white lily beds and sleeping river grasses that stretch away to the distant shores, where the sandpiper is flitting over the shallows. In the bright summertide he dreams and fishes—dreaming like the idle water beneath the burnished gladen. But as hunger sways him, you see him hovering like a gay little jewel or a blue pebble over the pure lilies, until his round eye, now no longer dreamy, espies a silvery fish, when he darts down, rippling the still water, and returns to his shaded perch to eat his little captive, head first. A tame little fellow he is, who has been known to alight on the knee of a silent broadsman and fish for water-beetles in the cool green shade of a reed boathouse, until an ejaculation of surprise sent him away shrieking, for he recognised his dread enemy, man. But tame though he be, he is unsociable; and if three or four birds get into a reed-bush together, they will endeavour to drive each other from the whispering car, unless they be pairs; then one couple will try and drive the other from the interior shade.

The kingfisher, I think, pairs for life, and indeed gunners seem to know this, for if they shoot one of a pair, they make sure to get the other before leaving—they must hang together. And what matter to them if their bodies do smell fishy; those whose educated sense is satisfied by a case of stuffed birds pay him well for his crime, and kingfishers grow yearly beautifully less—in truth, they are by no means common in the Broadlands. And their nests are rarely found thereabouts, they preferring some dark hole beneath tall trees hiding some small river to open lagoons.

In the prime of summer-time you may watch the gay little fellow fishing from your boat—catching prey all day, giving each fishlet a blow with his powerful bill against the side of the boat, then swallowing it scales and all. Upon one occasion one swallowed a small roach too large for him, and he flew in distress towards a neighbouring house-boat, the fish's tail protruding from his mouth, and plainly visible to the members of the household, who ran to his rescue; but on opening the window he flew away, frightened, to die among the willows perchance. But had he only known, he would have been safe; for the members of the house-boat had rescued one of his kind not long before—one who dashed himself against the glass and fell stunned to the deck, but recovered when placed by the fire, and flew off when permitted to fish again, the hero of a new experience.

Of this quaint bird many legends are told. One old Broadsman points to the brick wall of a decaying boat-house and says, "You see where that brick is all pecked away; them kingfishers did that. They neasted t'ree year running in a hole under that 'ere big gable." Another says, "Ay, lor', they're little warmin—bless their little bones; I have arn't many a half-pint on 'em. They prove knowing birds too— "They'll ketch fish and stow 'em up in rotten wood by the side of the dikes or river till they want 'm; they arn't all fool." Or another: "Bless 'em, they be clever at cotching fish, but I'm more cleverer than them, for I can cotch them." They are bright little birds, that add a bright note of tropical gorgeousness to our sombre grey winters, and outshine the brightest of our summer birds. Long may they fish by the pale lilies; long may they decorate the leafless willow boughs when the nor'-easter lays his numbing fingers upon the land, and makes one sigh for halcyon days gone, or only to be found across the grey seas.

YOUNG KINGFISHERS (*from life*).

CHAPTER LVI

THE CUCKOO

THE landscape is like unto a delicate pastel when the cuckoo appears in the Broadlands—soft masses of blue atmosphere, delicate patches of bursting leaves, long sweeps of tender green grass, a pale blue sky overhead, and the music of the warm breezes sighing over the face of the land. Upon such a day you may, on arising, hear the voice of the first cuckoo sounding from a coppice down by the water, and you listen as the echoing voice rings softly from the blue courting cuckoo, "cuckoo, cuckoo," who gives on this occasion two-and-thirty calls; and if you be Norfolk-born, you will know that you have two-and-thirty years to live—two-and-thirty more springs to hear the soft vibration of the cuckoo's voice. But the passing marshman will tell you you must wait for a purer note, for the bird hasn't sucked enough eggs yet to clear his voice; for he is said by all the country-folk to be hunting for eggs as he flies low over the marshland with that other Mephistophelian chortling chuckle of his, neither hawk-voice nor night-jar— a wicked, diabolical laugh, oftenest heard at eve and dewy morn, and never to be forgotten when once heard. But woe to you, says the rockstaff, should the cuckoo alight on a rotten bough, or, flying low, alight on a wall, for 'tis a sign of death, spring though it be; nor must you shoot him, or you will get upset.

Soon after the first cuckoo is heard, they are to be seen scattered over the marshland on the hedgerows, sitting on heaps of litter, airily perched on a fork-shaft; for the cuckoo dearly loves a pole or marshman's fork-shaft, or sporting on the river

walls, or in the cherry-blossomed gardens, hunting for eggs,
preferring those of robins, sedge-warblers, reed-buntings, and
titlarks, say the fenmen; indeed, they tell you he looks about
for eggs just like a boy, hopping from one place to another
for eggs to clear his throat.   And the old well-worn question
arises, Does he suck eggs ?   Throughout my sojourn in the
Broadlands I watched him, but never caught him yellow-
billed, though I have been within a yard of him all unknown
to him, and gazed in the bright sunshine upon his full bright
eye, sleek, whorish, greedy face, with feelings that he would
do anything that pleased him; for "duty" is a word to
cuckoos unknown.   But the evidence I have collected from
fenmen and others quite satisfies me that the cuckoo does
suck eggs; and though I never caught him, I have found eggs
sucked that were whole before the cuckoo hopped about them.
Moreover, I have met trustworthy men who affirm they have
caught him sucking thrushes', blackbirds', decoy ducks' and
reed-buntings' eggs.   One gunner, whose word is to be
relied upon, caught "Mr. Cuckoo" sucking a nest of decoy
ducks' eggs, put him up from the nest, and found one egg
still untouched, the rest being sucked, after the manner of
an old harrier.   Upon another occasion the same man was
huddled up in some stuff under an alder tree, watching for a
leveret, when suddenly a cuckoo alighted on the tree and
chortled.   Lying still as death, he watched the bird fly down
to the ground, when, to his surprise, he saw a duck sitting
very close.   When the cuckoo approached the nest chortling,
the duck got off her eggs and set at him.   The cuckoo came
on with his wings set out, his tail up, and his head extended,
making a hideous noise—"Enough to frighten a man," says
the gunner—"a loud, hoarse, hissing noise;" but just as the
fight began, in his excitement the man broke a twig, and the
noise alarmed the birds, who both flew off, the gunner finding eight eggs, "hard setting on."   Another fenman tells me
he has caught the cuckoo sucking mavishes' eggs.   Another
told me he saw a cuckoo flying with a thrush's egg still in

his bill. Another old man, a Broadsman, assures me they often suck reed-buntings' eggs, and he has seen them at it, and that they hunt the stuff regularly for them.

Finally, I have opened several cuckoos' crops at the beginning of the season, and have upon some occasions found a yellowish substance, which looked to me like nothing but egg. And there I must leave this much-vexed question, merely adding that I believe cuckoos do suck eggs, as do most predatory birds.

When the birds first come over, they hang about certain parts of the marshes; but in no case have I seen them take and keep stations, as has been said; indeed, they are constantly on the move, though they may feed about one place for some days; but so many mistakes are made by the inexperienced by mistaking hawks and night-jars for cuckoos, for their flight is similar.

After the first batch arrives, these strange birds keep coming over in relays, spreading over the face of the Broadlands, when they soon begin courting, chasing each other, flying low and straight over the green marshes and reed-beds, calling "cuckoo," and chortling or saying "cuck-cuck-cuckoo-oo," bounding up with the curious hissing noise spoken of by the gunner. And you may know their courting flights, for the titlarks join in the chase, but whether amicable or hostile I know not; sometimes their behaviour seems the one, sometimes they seem to cut at the cuckoo. The marshmen say they often hear the cuckoo talking to the titlarks and sedge-warblers, the birds answering them; and then, they say, they're on the look-out to suck their eggs and lay their own in the nest. "At this season," the fenmen say, "they are very attentive, along with the cuckoos." And I can corroborate their low hunting flights and apparent association.

And after the pairing comes the laying, the nests usually chosen in the Broads being the titlark's and sedge-warbler's, more rarely the robin's, wagtail's, and hedge-sparrow's nests,

but always the nest of an insectivorous bird, never the nest
of a corn-feeder.   It is stated with considerable emphasis in
most works on ornithology that the cuckoo frequently selects
the reed-warbler's nest ; this I believe to be a great mistake.
I have never seen a cuckoo's egg in a reed-warbler's nest,
nor have I ever met a fenman who has.   And when one
thinks of the structure and position of the reed-warbler's
nest, it seems very improbable that the cuckoo should lay
there, though he does, according to testimony, suck their
eggs.

Cuckoo's eggs differ greatly in size and colour, this difference being, as in many birds, due to the age and size of the
bird.   I think the stronger the bird the bigger the egg, for
the cuckoo is a long-lived bird ; indeed, I think one was kept
in a cage for over twenty years.   The cuckoo generally lays
one egg in a nest, though I have seen two in a nest, the
eggs being different, one being bigger and bluer than the
other ; but I think the eggs were laid by different birds.
One old fenman found six eggs in one season all in sedge-
warblers' nests ; one of these eggs was in an empty nest,
but after the sedge-warbler had laid two of her eggs, the
incurious marshman robbed both.

And when the young toad-like monster is hatched and the
foster-brothers are ejected from the nest, the busy little foster-
parents work hard all day to supply the ugly little beast with
food—beating the marshes for food, though they do not fulfil
the Norfolk superstition that the foster-parents go on feeding
the young interloper until they are themselves swallowed.
The young cuckoo leaves the nest before that, and ranges
silently, looking like a hawk ; for he is a quick grower that
wanders over the marshes, feeding upon caterpillars, stripping
the sallows bare of cankers, shaling round and round the green
islets of sallow ; and if you attempt to catch the youngster,
who is, after all, not a great eater, though a frequent feeder,
he will shriek like a hawk.   At this season, too, the youngsters are attended by a retinue of birds.   I have often

watched ten or a dozen little marsh birds buzzing round a young cuckoo. According to local tradition, most marsh-birds will feed the young cuckoo as if he were the king of birds, but I have never observed this fact. And when the polygamous males have made enough love, and the careless females have deposited enough eggs, for each bird lays several eggs, to judge by the survivors, they go flying about the marshes, the males whistling and babbling through their quiet, strange existence. At this season you may see them hunting for caterpillars as late as nine or half-past nine of the evening, when the grasshopper warblers are grinding and the partridges calling on the uplands; and well may they whistle, for they have no cares, never feeding their young, or indeed keeping in their neighbour-hood. And at the end of July they go, leaving the dark silent youngsters to the mercy of their foster-parents and friends, finally, in their turn, to find their way across the seas alone, unless they are born " with nawigation," as the fenmen say, for they are regular " chummys " of the marsh-mowers, sitting on their forks, upon osier-stubbs—a favourite seat—or upon a heap of newly-poled stuff, sorting their feathers and uttering "cuckoo," sometimes until they are quite hoarse ; or, if disturbed, they will gurgle or chuckle, or they will fly off and alight with spread wings and upturned tail on the wall, calling " cu-cu-cu-cu-cuckoo" or "coo-cuck-oo " —another bird answering across the green sea of marsh-land; and chummy though he be, if he fly along the white road, winding away amongst the green hedgerows, ahead of the home-returning marsh-mower, the fenman is sad and solemn, for it's a rockstaff that there will be a death in his family.

And in September, when the fenmen tread the purple loosestrife, and the upland and the marshland are turning sere and yellow and grey with the colour of decay, the young cuckoos start without compass or leader across the unknown sea, leaving their parents, the affectionate little titlarks or sedge-warblers, to mourn for them; for have they not fed them from toad-like childhood until they were as large as their parents? Indeed, the little birds love them, and I always think that chasing of the old cuckoos in early spring may be the foster-parents looking eagerly for their child reared the summer before on the grassy marsh, or down by the clear watered dike where the young pike now darts to and fro. Strange whorish character that he is, the cuckoo's habits still remain a mystery to all.

And after all this observation of the cuckoo's habits, the only lasting impression we have is that of a family of gay careless chortling heteræ, coming out of the deep and flying from the blue misty coverts over the marshes in the prime of summer-time, feeding on all the delicacies of the season, undertaking no work or cares, imposing their duties upon others, deserting their own offspring and leaving them as foundlings in the nests of respectable little birds, whose hearts they may break ere they, too, are lost again in the haze overhanging the grey sea, and then throughout the dark and dreary winter watches all that is left is the echo of their joyous calls in the heyday of spring-time, when all the world is young, and every lad hugs his lass.

CHAPTER LVII

THE OWLS

ALL gloom-loving birds have always the mystery of crepuscule attached to them; their very distinction is due to their uncommon feeding hour, grotesque appearance, and solemn vacuity; for, far from being a bird of wisdom, the owl is, like many a solemn savage, a low type intellectually, solemnity and mediocrity being oft found together, two qualities that impose upon the superficial observer merely.

Three kinds frequent the broadlands—the barn or church owl, the long-eared or horn-owl, and the little short-eared or "marsh-owl," as he is commonly called in Norfolk.

THE BARN-OWL.

When the silvery mists are rising and lurking on marsh and mere like phantoms, you may hear the shrill screech of the barn-owl as he beats with heavy flight round the stacks and out-buildings of some lone marsh-farm in search of mice or young rats—his staple foods.

At all seasons of the year you may see this bird hawking through the gloom, and if you be an expert with your lips—shutting them tightly, and suddenly drawing your breath in, making a squeaking kind of noise—you may decoy him to fly between you and the stars; but his large eyes detect the fraudulent imitation, and he flies contemptuously overhead to some big ivy-green tree or stack to meditate on that absurd creature man; but that is one way to shoot him. Another, practised upon "softs," is to give them a sieve,

and get them to stand in the soft light of the moon under
a stack, watching intently for an owl to alight on the
yellow-grey straw.   As he stands intently looking for the
wary bird, a pail of icy water is dashed upon him from
behind; and even at sea some "soft" will look for the
owl down the galley-chimney, receiving a ducking for his
pains.

Perhaps the stupidity of the bird when startled from some
gorse or reed-covered shed into bright daylight accounts
for the contempt in which the owl is usually held by the
fenmen, for he flies stupidly about till he finds another
shed wherein to lie up.   In the spring the boys search the
steeple or ivied ruin or tree for his well-known eggs and
young; but they must take care, for he is a dangerous
fellow when attacked, as a friend of mine found.   He shot
at an owl one evening on a marsh near a farm, and winged
it, the big bird fluttering to the marsh.   Taking off his
cap, he ran up to capture him, when the big bird drew
back its head, its great eyes blazing with fury, opening
its mouth and hissing like a serpent.   As he approached
the irate creature, the owl seized his cap in its talons with a
vicious and exultant grip—a sort of "I've-got-you-now"
feeling gleaming in its fierce round eyes.   Whereupon
my friend began the battle by seizing its throat, trying to
strangle him; but he could not, and at length he felt his grip
loosening, and the big eyes blazed fiercer than ever, when
suddenly it let go of the cap and seized him by the thumb
with one talon, the sharp sickle-like claws meeting in his
flesh; and, indeed, the owl now had the advantage, for
my friend's grasp upon its throat was getting weaker and
weaker, and the pain severer, whilst the owl was getting
more vigorous than ever.

As my friend looked helplessly about him, he saw the
sluggish dike, and carrying the bird over to the water, he
knelt down by the grassy shore and plunged his hand, owl
and all, into the water.   A few bubbles arose on the still

surface of the dike, and all was over; nor does my friend wish to fight any more owl. He says the brute was most savage and terrible-looking; and what a poor mouse must feel when he sees that great creature pounce upon him must be left to the imagination.

THE LONG-EARED OR HORNED OWL

is to be found in the plantings and cars by the broad-sides; but he is now rarely to be seen, although once I saw one on a gate-post in the high noontide. But his curious voice can be heard for a mile. An old keeper, who once had charge of some coppices abutting on a broad, told me this bird is fonder of the sparrow than anything else, and that he once found nine sparrows with their heads off near his nest. He says they lay five eggs when young, but only three later on, generally choosing an old pigeon's nest to lay in at the end of May. One of these birds once took a nest full of young swallows from the eaves of his cottage.

THE SHORT-EARED OR MARSH-OWL,

as the fenmen call him, or the "little horned owl," as others call him, is sometimes flushed in the shooting season, when he is mistaken by the inexperienced for a "woodcock." But from the few specimens I have seen, it seems to me, in a fair light, this mistake is inexcusable; but then I was not after woodcock, and so had not woodcock on the brain. This owl has the flight of a kittiwake; the woodcock has not. But I have never seen many, and never yet a nest. The last nest I knew of taken in the district was in the year 1882, therein being six eggs. A friend, a fenman, who found another nest in 1878, gives the following account. He has, moreover, taken several "mash-owl's neasts" in his time. He said :—

"I was mash-mowing 'bout a quarter of a mile off the

neast in a dry mash, when I see the two old buds going back'ards and for'ards in the daytime, beating 'bout the mashes like a harrier. I know'd 'em in a minute by their fly. I watched 'em; when one on 'em caught a mouse or a bud he would rise up and soar thirty or forty yards high afore the wind, then off he'd go to his neast. I watched 'em two or tree days at this game afore I went to hev a look; there they was huntin' like an old buzzard just above the rushes, and all at onest one would dart down and catch his grub. They don't huver like a kestrel. I saw the male bud did most of the catching; he kept going off and droppin' onter his neast. Since both of 'em was hunting, I knowed there must be flappers; so off I go ontil my face was close; and as I drawed nigh the neast, the old cock bud kept calling overhead, and when I got right agin the neast, the old hen flew up and flew on a few yards, and then she went tumbling on ter mash, as if all her bones was broke, she was that artful. If I'd been fule enough to follow her, she'd have gone on for ever. I follered one onest; she led me acrost tew mashes. Time she was doing this the old cock was hanging shriekin' over us. There was the neast, just like rushes beat down; there was tree young 'uns and one rotten egg; they were in down. There was a lot of mice and birds at the neast, and a young half-grown water-rat (vole). I took the egg and one young 'un, and I took him home and fed him on buds, mice, and rats; he would eat nigh anything. He held 'em in his claws and tore 'em up. When he was wery young he was wery fond of turning onter his back and shriekin'. He got wery tame, wery tame indeed. He got so he would watch for me coming home; he fared ter know when my time was. When he see me he would come with his wings out a-shriekin'. But he hev mistook other folk for me. I sold him to a gent when he got big.

"I hev killed lots on 'em. I hev put six up off a mash —not all at onest, mind you. They was a clutch, no doubt.

If you see one of a winter-time, and make a noise like a mouse, they'll come and hang a yard or two right over your head.

"They set on the mash in the rushes of a daytime, and lie very close when they ain't hunting; I have put 'em up in July: full-flappers. If an old rook see him of a daytime, he'll trosh him in the air; but once he get onter the ground, they daresen't interfere with him. I have often seen him hunting tree o'clock of an aternoon; they don't mind daylight no more than you nor I. They roost all in a clutch back of the lumps of stuff.

"They're little warmin to fight tew. I winged one t' winter, and sent the dorg arter it. But he tarned on his back, and caught the dorg a shivering tip on the face, and old Jocko wouldn't face him no more—he had got enough, I reckon."

The owls then do not give us much opportunity for study—a screech, a hoot, a snore, a barking or mewing, a sucking sound, a noiseless bird flying through the gloom or over the moonlit marshes, a fluttered object chased by small birds in the daylight, a solemn creature roosting in a dark shed, by accident discovered. Quantities of casts, little parcels of fur, feathers, or scales and bones—mingled remains of rats, mice, water-voles, sparrows, and other small birds—these are merely the signs of the owl, the evidences of his existence. But his real life is but little known, nor can it be anything else, for he is a night-bird.

But, mysterious as he is to us, he is a good friend—one of the farmer's best friends. Never does he harm the game-preserve; and yet the bumptious, bragging keeper—that flunkey of much knowledge and much ignorance curiously interwoven—will slay him ruthlessly, as he does the kestrel, and the reward is a plague of mice, that eat the farmer's corn, and there is no gold to pay the landlord, who hires the keeper who kills the owls, who kill the rats and mice that eat the corn that enrich the farmer, who pays the

gold to the landlord's keeper who kills the owls.   And so
on, and so on.

The Wild-Birds Protection Act wants thorough over-
hauling.   Some birds now protected should be struck off the
list, and other birds should be protected all the year round
under heavy penalty, and amongst them the kestrels and
owls.

CHAPTER LVIII

THE HARRIERS

THE MARSH-HARRIER

STRANGE active birds of the fenland, half-hawk and half-owl, they come and go, and at every occultation cause sorrow to the gunner's heart, for there is a price upon their heads, and I don't know that their extinction will harm any one.

On May-day, when the sallows are covered with leaves, and fresh green islets of covert rise from the grassy seas, an unmated male marsh-harrier with cream-coloured head may appear, and be seen beating to leeward over the soft marshes, rich with soft rushes, sedge, and scattered reed; for they love the water, and perhaps the rail's and waterhen's eggs and young they find thereabouts; for such a marsh as the marsh-harrier loves, the rail and moorhen love.

Directly the light breaks in the northern sky, you may see this lean bandit, all bone and feathers, with his slow, heron-like flight, beating over the green reed grounds and shimmering beds of floating gladen on the hunt for the spotted eggs of rail, waterhen, coot, or young snipe, which he dearly loves. As you watch him in the growing light, perhaps you may see him fly down into a gladen bed and disappear; and in the cool fresh morning you will hear the heart-rending cries of the mother waterhen, as he sits on her nest and deliberately makes his breakfast of fresh eggs or dainty young black water-chicks. Presently he rises lazily; and as he flies up over the stuff and goes on his way, chased by a cock peewit, whose wife is sitting near-by, perhaps a lurking gunner drops him, and as he picks him up, the yellow yolk and thick bits of shell run from his guilty mouth to the marsh.

As the sun rises and disperses the mists rising from the resting waters of the mere, and the day brightens, the sky turning blue and the waters on the land glittering, he flies off to his moist morass, and sleeps through the bright heat of the day, for sunlight is hateful to him, thief that he is. He does not like to "work" in the high noontide.

But when the sun begins to sink behind the reed-beds to the westward, he leaves his watery lair, and is often to be seen shaling, swimming round and round in the azure, rising, beating to windward—beating his long lithe wings a few times as he comes to the wind—then, turning lazily and rising up, another turn of the aërial screw, without any apparent effort; and when he has played in the air to his sweet content, he descends from his aërial flight, and uttering a few sharp cries like a kittiwake, he begins beating his old station over the swamp crops in search of young birds and eggs ; but eggs he loves best I think, and a good taste he has in eggs too.   But he never takes birds on the wing like a hawk.   For a fortnight or three weeks you may see him thus living a predatory life in lonely bachelorhood, when one fine morning you see his hen has arrived from some strange land, and both go beating the broads.   I have known them go ten times a day round the very same beat round a broad-edge ; indeed, this is characteristic of the hawk and harrier and owl tribe; and if perchance they drop upon any prey, and whilst eating it hear a shot, they fly off, soon to return to their treasure to finish it.

But soon love inflames the cock's bosom, and he begins, after the strange fashion of many birds, to build "cock's-nests" upon some little eminence on a soft marsh rich with a crop of swamp-grass, for a bare marsh is hateful to him, either for hunting, sleeping, or nesting. You will see large pieces of thistle-stalk and sedge piled carelessly about. Thus silently exhorted, the darker, bigger, more richly feathered female begins to build in earnest, choosing a little hill amongst the rush, or, if none such is to be found, piling high the stuff, so the wet shall not penetrate to the eggs. And once she begins her nest, at the end of June or beginning of July, he falls to and helps her—in fact, doing the lion's share— piling high the sedge, rush, and thistle-stalk, till the littered pile is pronounced finished. Indeed, she does little work at any time, rarely "working" more than twice or thrice a day. And when the first hen-like egg is laid, she begins to sit at once, and very closely she sits; you might pass within a yard, but she'll not heed you, nor will he, if he be near.

Again, when the four young ones are hatched, he does the catering, beating over the warrens and marshes for young rabbits and leverets, which he brings and drops on a heap of stuff, hovering some twenty yards above the ground as he drops the food on the eminence; and she, when hungry, comes off her nest and eats her meal leisurely on the heap, casting up any indigestible pellets of bone or fur.

And when the young are strong on the wing, you see them beating a few yards above the reeds, hunting for late eggs, or else feeding upon big dragonflies, a favourite food with them when the egg season is past.

A strange irony of fate is this by which a "tom-breeze" or dragonfly himself hawks through a cloud of midges at sunset, snapping them up as he flies through them like a swallow, and then is himself snapped up by a hungry marsh-harrier as the pale moon is rising in the sky; for the birds work till nearly dusk in the late summer evenings; and if you disturb any of the family, they'll go shaling up into the greying blue, round and round, sometimes shrieking like kittiwakes over the green sea of rush; for the rush marsh is dearest to them; indeed, their nest, as often as not, is merely

a heap of broken-down rush and green reed, often having all
the appearance of being made by the birds dropping from
a height into the stuff and breaking it down into a formless
mass.

The last nest I knew taken was in 1874 by the fenman
who found it.   He said: " 'Twor in July, we was mowing
rushes in a rush mash, when we see the old bird hovering
right over the top on us, but we didn't pay no regard to it,
though that kep on shrieking like an old kitty down on the
North Sea.   We kep a mowing, and all at onest I mowed
as close to her as I am ter you, and up she jumped, and I
picked over the stuff and there lay her neast, built o' rushes
and reed, wid four egg in her.   I let the eggs be, and runned
for my old betsy, but twarn't no good; up she go shaling
right out of sight."

At times the young birds are to be seen sitting silently on
marsh posts, heaps of litter, and on the walls—that familiar
dining-table of the marsh-hawks, owls, harriers, and even
herons, as their plentiful casts indubitably prove.

But few of them escape the ubiquitous gunner after shooting comes in, and one old gunner assures me he shot one
in November, and I believe him.   I have seen them in
September, and farmers tell me they have met them coming
sharply round a stack after harvest, as if they had been
hunting for something thereabouts.

But few go away at all—the blood-money makes them
eagerly sought after by every gunner in the parish; and
perhaps they know this, for I have not heard of a nest being
found during the last few years, though I have seen more
than one pair, and sought their nest for many a day round a
sallow-bush—a spot sometimes chosen by them for nesting
—yet no nest could I find, only dung; it was evidently
their roosting-place, if nothing more.   Young birds, however, are seen about every year; so, after all, it is probable
they still breed in the district.

THE HEN-HARRIER.

" I have shot female hen-harriers every month during the winter," said one of the best professional gunners in the Broad district to me one day; " but," he continued, " I never saw the male bird here at Christmas-time, only after February or March, and oftenest of a summer-time."

The female hen-harriers are not so very rare, and in the dark days of winter you may often see them beating some ten feet above the reed-beds and marshes, with heavier and more laboured and more frequent beats of the wings than the Montagu harrier; they are easily recognised when once seen, though they sometimes hover like a kestrel, even in a blinding snowstorm. I saw one of these birds, one early spring, hunting unconcernedly over a yellow snow-spangled reed-bed, wandering away, a dark speck in the snow, fading gradually down the powdery landscape over the wide marshes; for they go farther afield than the marsh-harrier, and beat a greater round in their hunts for eggs, mice, young birds, young rabbits and hares; but young snipe are their favourite food. But their usual beat in hard weather is the rond, where the rails and waterhens and snipe congregate by the little trickling rills left by the tide or melting ice. On a hard, bright, east-windy day, when the reed is yellow and the sky blue, and the dikes and rivers and ronds frozen hard, you may see an old female hen-harrier beating heavily along the rond, and suddenly stop and hover for a moment, like a kestrel over a mouse, ere she darts down and seizes a shrieking waterhen, taking her to the hard frozen river-wall to devour hungrily in the keen air. And when hunger, that impels all beings to extreme measures, shall have gnawed at her empty stomach, and the rails and waterhens are scarce, she has been known to hunt a full-sized hare, flying just above him as the terror-stricken wretch ran along the frozen wall; but the fenman disturbed the sport, and never saw the end.

M

But surely there can be little doubt but that the hare was mastered in the end by the fierce and hungry bird. And this may account for many a hare's skull I have seen bleaching by the waterside, its hollow sockets staring into the blue with an empty melancholy.

A bold bird, too, is the hen-harrier; for at times in the dark winter afternoons the farmer will be startled by her coming round his straw or corn stack with a sharp turn, and if he chance to have his gun handy he kills her.

Formerly they bred on the marshes, but I have heard of no nest being taken since one found by a fenman well known to me in the year 1870.

Both birds are to be seen even now. Only last year (1891) a fenman shot a cock-bird sitting on a wall in June; but the price set upon their heads is too great, and they do not remain long unshot, and are so prevented from breeding. Perhaps they do not escape destruction so often as the marsh-harrier and Montagu, because they like the dry marshes nearer the farms and villages, and are therefore more easily to be got at. Of one thing I am certain, however; were no birds allowed to be shot in the spring and summer, the hen-harrier would again nest amongst us. As it is, the last that I know of was found some fourteen years ago, and this is what the finder says :—

"We didn't know noathin' 'bout the walue on 'em then. If we shot one, we used ter send him ter market, and get half-a-crown for him, same as other hawks. But I found out arterwards they was worth money. 'Twas this way. I and my old chap was a-mowing of rushes on a dry mash agin the mill. In the course of the forenoon we heard a bud shrucking over our heads. We didn't think 'bout no neast, so we went on mowing, and I mowed onter it. I mowed as nigh to him as the middle of the next swathe, and there lay five eggs in a neast made of rushes and short reeds all broke down like a swan's neast. So we must leave

'em that night, 'cause we hoped to get the old bud and all in the mornin'.

"So next mornin', directly that was light, I went to look if she was on. Time I was peeking for her, up she go, and I slung my flail basket-stick at her, and hit her athwart the legs; but that didn't stop her, no jolly fear. She cleared right off. So I must take the eggs, and sold 'em to Josh for half-a-crown. I didn't know they was worth a lot, but I soon found out. For a few days arter along comes a gent as collected eggs, and he wanted ter know all about it. So I told him, and showed him the place; and then he went to Josh, but he had sold 'em to Norwich. So away this gent go to Norwich, and wants to buy one egg. But, ' No,' says the chap, ' I won't break the clutch for five pound.'

"And that come to our ears. So we got ter know they was worth money; and aterwards we used to get a pound for the old buds, but I ain't seen none lay since that—the chaps is all arter them, and won't let 'em lay, else they'd do so right enow."

It is said the harriers have left the low grey fens on account of the drainage. I do not think this is the case, but submit that they left because the coots, rails, and water-hens' eggs have become so scarce; for they love an egg or young waterhen, coot, or rail, and these last few years these have all become rare. Water-rails, since the winter of 1890–91, are more rare than spotted rails used to be.

THE MONTAGU HARRIER.

When the hawthorns are decked with flowery petals and the marshlands are a field of cloth of golden crowsfoot, and the sallows are leafy with June foliage, the cock Montagu, or "Blue Jacket," arrives on his favourite soft, moist, rushy marshes, where the ragged-robin blooms on the higher islets, and the kingcups blaze from the bistre marsh-water; and where he frequents there they will build; indeed, he seems to come over of set purpose to build, and often begins to build a rough nest of sedge and soft rushes, and the pale grasses that grow like the stately rush.

In the beginning of this leafy month, unless the season be very backward, you may see old cocks here and there scattered over the rush-marshes, sitting round on lumps of stuff by the hour together, waiting for the females, who arrive a fortnight later.

Early in the morning, too, when the mists rise from the cooling marshes, heated by the previous day's sun, you may see one hunting very low, some two or three yards above the reeds and swamp crops; at times darting down like lightning upon a mouse, nest of waterhen or rail; then up he gets again, flying backwards and forwards over his regular beat, which rarely extends more than a mile from his flat cock's nest. He seems a quicker and shyer bird than the other birds—not so venturesome. And at this season the ubiquitous gunner is upon his track, hiding in a clump of sallow or behind a heap of mown litter lying in his beat, watching for the big bird to loom through the morning mists, and as he flies overhead, the tongue of fire shoots forth after him.

But soon after they have paired the nest occupies all their time that they are not feeding, for the cock's first attempts are either ignored or improved upon by both the savage housekeepers. If the marsh be moist, the flat nest—smaller than a marsh-harrier's—is raised from seven to fifteen inches from the marsh bottom; on the other hand, if the marsh be dry, the nest does not rise much above the ground. And the materials vary according to the marsh crops growing alongside—old sallow sticks, grass, soft rushes, sedge, and occasionally a few of their own feathers being the chief stuffs employed. And directly the first egg is laid on the reedy boat—floating as it were on the green sea—the hen begins to sit, and closely she sits, never leaving the nest for long. Indeed, many fenmen have nearly caught her with their hands whilst sitting, so devoted is she to her four bluish-white eggs.

And some fine morning, if you know the solitary islet near which or under which the nest is piled up, you may, if lucky, see the cock come out of the blue with a nearly devoured morsel in his mouth—either a lark picked to pieces, a water-rail or moorhen, or, more rarely, a young rabbit or leveret, and hover some fifty yards above the nest, calling with his wild voice; and the hen flies up into the blue to meet him, taking her breakfast with a shrill cry, flying off sluggishly to a hill or some heap of stuff to eat it on an eminence whence she can see the creeping gunner, whom she dreads as the lark dreads her. There, too, you will find those strange pellets of hawked-up bones, fur, and feathers. But sometimes the gunner is too artful; then a wild shriek resounds over the muffled marshes,

HAUNT OF MONTAGU HARRIER.

fading away in the mists, and the mangled cock tumbles to the marsh, or flaps on to the marsh-wall or some neighbouring heap of stuff, where he begins incautiously to dress his wounds; but the stealthy gunner has followed him up again and pours the lead into him, and he falls dying upon the damp ground or litter; and as the keen gunner breaks forth and seizes him, tossing his long, limp, thin body against the blue sky, the yolk of stolen eggs streams from his mouth, and he is convicted of theft.

In early spring perhaps, some fine morning, you will not see a cock Montagu in the sky, when suddenly a brown hen flies with her heavier beat in from the sea, and then the blue air resounds with a far-reaching kittiwake-like shriek. The shaling cock has seen her, and flies down like lightning to court her, and perhaps to fight another cock, who has been waiting for the hens as well as he, for there are generally more cocks come over than hens; and they fight fiercely, as the fenmen bear testimony, though I have never seen one of these love-combats, but fenmen tell me they have often seen them fighting and shrieking in the air at the pairing season. Then, too, the fenmen are on the alert, hiding up to watch for them; and if ever the birds catch a glimpse of the gun-barrel, they begin to jerk themselves about, and the fenman waits patiently till one steadies himself again, when he " cuts it inter him."

The more wily gunners, too, leave rails' and waterhens' eggs in their beats as a decoy, hiding near by, and when the birds pounce down to suck them, they fire upon them from their ambush.

And if you wander through the swamps, raising a cloud of midges that sting you into red lumps, the cock will sight you at once, and come towards you, like the swan, full of inquiry, turning to right and to left, and shrieking his wild call; then you see him go up into the blue some two hundred yards, then turn on his side and fly about like a snipe, going more quickly than before; but if you disregard these antics, aimed at leading you away—for if you follow him he will take you a quarter of a mile in one of these ascents before you know it—and look for the nest, you may presently step right on the hen-bird crouching low like a pheasant on her brood, when up she starts with a shriek ; and as you bend over the nestlings, the old birds return and fly about in the liquid air above you, shrieking their odd notes and fluttering down, almost touching your cap; and when the young are robbed, they follow crying, for a short distance, then turn and disappear in the blue, leaving their white fluffy children in your possession.

After very hard winters, such as we have had of late, they do not seem to breed; the leafage hangs back on reluctant wings, and they seem to go on elsewhere to nest.

"And when do they leave us?" you may ask; to which I answer you as the gunner answered me to whom I put this question—"I dunno when they go away; I never see'd 'un go away." He himself had killed eight birds in one season, and, truth to tell, few escape the long ten-bores, whose owners are always in search for them.

MONTAGU HARRIER'S NEST AND YOUNG. (*From life.*)

CHAPTER LIX

THE COMMON BUZZARD

THE fenman calls the buzzards and harriers indiscriminately "old buzzards," and in winter-time every man who owns an old muzzle-loader turns out as soon as he gets word that an old buzzard has been seen shaling about the marshes.

The common buzzard now rarely appears lolling along over the dry sandy warrens, where the silver-weed and rag-wort grow on the poor soil, in search of an unsuspecting rabbit, his favourite dish.

I have once or twice seen this large gloomy-looking bird flying toward a reed-bush at eventide, for they prefer to sleep in the stuff, it being warmer there. You can always tell them by their flight, for they fly more swiftly than a harrier, and do not "lop" so much, as the fenmen say; in fact, they have not to hunt the thick stuff closely for a living, as does the harrier, although the old buzzard has been seen to take waterhens, coots, and partridges, and in early spring young hares.

When hard pressed by keen weather they will eat most living things that cross their path. In spring and summer they rarely frequent the marshland, and never breed, though the buzzard generally leaves in March.

He, too, is a mere passing vision, a brown patch to be seen now and then under the grey winter sky, ranging over warrens and marshland—a reminder of days gone by —of days when a different climate obtained over these once solitary wastes.

CHAPTER LX

THE SEA-EAGLE

ONE lovely day in the early spring of '91, as I walked under the tall steeple of Winterton Church, I saw a large bird beating over the hummocky warrens, flying like a harrier.

The sandhills glittered in the sun like snow-peaks, and the bleak scrubby vegetation of the sandy marshes around that squalid village looked blacker than usual as I saw this large bird alight on the warren to eat some incautious rabbit. I suspected him to be an eagle, and my suspicions were confirmed; for that very night a keeper at Somerton shot him as he was flying into a fir-tree in a planting to roost for the night. And such are the glimpses you get of the eagle in the Broadlands. A large bird is seen shaling above the marshes or "warrants," and you hear a day or so after that an "eagle" has been shot. But the birds are very rare thereabouts.

A young fenman, in whom I have complete trust, told me that when he was a boy he saw two of them beating over the "warrants," like great old "buzzards," one day; but the next, he came across one of them imprisoned in a trap, and the remains of thirteen rabbits around him, food that his mate had brought him.

We merely get glimpses of this eagle in the Broads, and I have never heard that wild saddle-back gull-like cry of his, as I have elsewhere.

CHAPTER LXI

THE HAWKS

THE "GAME-HAWK"

ONE spring, as I walked inside the sandhills in the roar of the sea, two powerful birds of the hawk tribe flew swiftly overhead. "Game-hawks," said the fenman at my side. They resembled a sparrow-hawk in build, but were altogether a bigger and more powerful bird; they may have been falcons, but I could not be sure. I hesitate to say anything about them, for they are birds seen rarely, and then only for a moment perhaps. But the fenman seemed to know them well. They were common in his young days, he said; and he had often seen them "go cruising about, and if they chance on a bunch of peeweeps, they'll pick one flying by himself on the outside of the bunch, and they'll go arter him, and cut him down clean like a clod. Ay, and they'll go at old mallard and coots tew. I never seed 'em arter small bahds. They're rare rum bahds. When they fly, they don't move their wings much—kind of slade along, not like them rough-legged falconers *—they go lolling along. And they're wonderful fond of rabbits. They'll go and pick the young 'uns up in their claws, and you can see 'em flying low along the warrents, everything flying up ahead on 'em, peeweeps and all."

The birds I saw flew exactly as he had said, easily and swiftly, sliding along, as he would have said, into the distant blue, over the marram-crested dunes.

* Most likely the rough-legged buzzard.

THE SPARROW-HAWK.

As you sail near a dripping plantation in autumn or early spring with a full breeze in your sail, you may see the air full of flocks of birds driven across the sky by the wind from the fields near-by. Still you glide on, when suddenly you see a commotion in a flock of sparrows or thrushes, as they go flying from the wood across the river, and you look up near the tree tops, behind which the yellow sun hangs like a ball of fire, and you recognise the sparrow-hawk hunting for his dinner: perhaps a young bird who has never left the fields near his native planting, where he was reared in a nest of sticks from the great blotched white eggs nursed by his mother. At another time you may creep along a hedgerow in winter, and hiding in the close ditch, watch the sparrow-hawk hunting along the other side of the hedge. He knows where you are hiding, and is, like you, watching those hungry flocks of larks, thrushes, sparrows, greenfinches, and chaffinches feeding on the stubble; but he will fly leisurely just above the ground till he is safe out of your way, when suddenly you will see him wheel—for he is a swift flyer, resembling a pigeon on the wing—and dart through the hedge, seizing an unsuspecting chaffinch, sitting quietly down to eat him where he catches him, the rest of the birds rising in a noisy flock and scattering like chaff before the wind, seeking the nearest hedge, for once in there they know the sparrow-hawk cannot follow.

In very cold weather, when the dikes are laid, and in the marsh-farm the cattle are in the yard, he hunts about the straw-stacks for his breakfast of sparrow or greenfinch, or, failing that, he goes down to the marshes and hunts for larks, his standing dish.

All the year round he is not averse from young mavises, blackbirds, and partridges, when he can feed upon them on the arable marshes or grassy lokes. And if you see him

catch a bird, and watch him, you will find he oftener than
not eats it where it died, afterwards flying off to some
post or pile of litter, or the raised marsh-wall, to digest his
dinner; but more generally to a post, where he sits looking
sleepily at you, sluggish as an alderman after a city dinner.
But he is not so common with us, and, lover of the coppice
that he is, he is not a great frequenter of the Broadlands.

THE MERLIN.

The little "blue hawk," as the marshmen call him, or the
"sparrow-hawk," as others call him, is rare, though I have
seen him dart over the marshes in winter like a rocket
striking at a red-poll, for they kill on the wing; but 'tis
a mere meteor-like vision, which makes me realise a story
told me by an old bird-catcher of how his brace bird, a
goldfinch, had its head taken clean off by a merlin on the
wing. Truly a blue and destructive comet of the marshland
is the little merlin.

YOUNG MERLINS. (*From life.*)

THE KESTREL.

The kestrel is the watchman of the marshland. At all seasons of the year you may see him hovering over his prey—either when the broads are white levels, or when the marshes are gay with many-coloured flowers, or sere with dying grasses, or fresh with new spring grass. He is the true mouse - hunter, and the long - tailed marsh mouse is his ever-ready prey; indeed, every marsh-wall is strewn with their soft mouse-coloured down edged with brown, for he, like all the hawk tribe, loves to eat his food securely on an eminence, whence he can cast his keen regards over the flat-land, and detect any coming danger. Freshly picked mouse-fur and casts, these are unfailing signs of his dining-places, where he spews after he has pulled the fur from his little bead-eyed victim, and rent him with his bill, and eaten him from the head downwards. But if the weather be hard, he falls upon larks, sparrows, water-voles, young rats, and, in the breeding season, frogs and young birds, even young partridges. But he takes few of these—very few; and, everything considered, his name should be erased from any list of outlaws, for he is one of the most useful birds we have—a real farmer's friend, suppressing the plague of field-mice with a firm and deadly claw.

But let us watch him just before closing-in time on a calm evening, when the marshland wears a bland aspect, and the light is subdued and diffused over reed-bed and gladen morass—for he prefers the evening and morning to eat in, though he may be seen hunting at every hour of the day. Then he comes flying lightly down his regular beat by the marsh - wall leading from the mill to the river, and he suddenly stops, his body well held up in the air, but his head kept down, his tail spread fanwise and bent down, and his wings flapping quickly. He is perfectly balanced

—treading air, so to speak—there being no slipping forward, as has been erroneously stated. He keeps in this position until he espies a mouse, when he closes his wings and tail, and drops through the air like a stone. See, he has got the marsh mouse, and is going to the wall to eat it, as there is no post handy, for he prefers a decaying post.

Or you may watch him on a windy day, when the marshlands are ravished of all delicate harmonies of colour, and everything is spotty and noisy. Then you will see his body lies straight on the wind, his tail vanes are closed, and it lies out straight behind him; whilst his wings are spread, merely oscillating sufficiently to keep his balance. He is floating over his prey now, and he will either rise higher in the aërial sea, or drop deeper, according to the growth on the marsh or the light, for he must focus his prey before he seizes it.

On these days of the breezy marshland he loves to sport—as indeed do golden plover and peewits—mounting up through the aërial sea, always facing the wind when he begins to rise, starting from a spring-cushion, as it were, and flying round in a spiral—but no mere erect spiral, but a spiral that is always working to leeward. So you may watch him mounting up, giving three or four quick flaps with his wings as he ascends to windward; for by the flapping is gained impetus sufficient to carry him up the next incline of the spiral, then out go his wings, and he turns on his side, and up he goes for a considerable height, but always well within sight.

I never yet saw a kestrel " soar " straight up without moving his wings, or go very far up into the blue, as has been asserted, and I don't believe he does; this, like many of the late Richard Jefferies' natural history notes, is inaccurate.

And you may watch him hawking over the marshlands. If one old bandit meets another, either one will give up his hovering ground and fly off, or else they will fight right merrily with claw and beak; and if you disturb them they will often go up into the air in the way I have described. I have seen a young bird who was hovering over a favourite pasture fly away immediately an older bird came up, the elder taking up exactly the same position in the air, hovering over the very same spot on the marsh, soon making a kill; but his prey gave him a little trouble, for he hovered thrice before he made a kill, for it is their custom to hover every time their prey goes out of sight. This practice has been mistaken as part of the mechanism of hovering: it is not so.

The kestrel will take a beat for days and days, when perhaps he will change, going another beat; but he will keep about the same marshes for weeks and weeks.

And in the spring-time you may come upon them courting; indeed, I think they make love in the air. I have more than once come upon a pair in the spring-time, the male holding the female by his claws, both being about thirty feet from the ground, and when they saw me they began to descend with shrieks. Their feathers and wings were ruffled, and on touching the ground they parted, the female flying off across the marshes, the cock-bird soon following her, shrieking as he flew.

And about the end of May they either take an old rook's nest or select some dark hole in a water-mill or chimney, or a church steeple, where they lay their reddish brown eggs and hatch their pretty, fierce-looking youngsters. One pair has for four years laid in the same wooden box at the head of a little skeleton-mill, though they have been robbed every season. The last clutch I saw taken from this mill was on a bright day—the first of July; their big, clear, bold eyes, with their yellow waxlike rims, staring with dignified contempt upon us. They had built no nest, the box bottom serving their purpose. Thence we took the skeleton of a frog and some bird's feathers we could not identify, besides several pieces of field mice.

One fenman I know tamed one of these fierce-looking

youngsters, feeding him upon young birds, water-rats, and dead rats, all of which he ate greedily; and if he tried to take a bit of food from his yellow clutch, the little " varmin " would turn upon his back and fight with his feet, shrieking all the time. At last he got so tame that he would fly shrieking down the road to meet his master, and welcome him home.

During the nesting season they are scarce on the marshes, and I have passed some days without seeing a bird; but directly the young leave the nest the family takes to the marshes, often following a solitary marsh-mower as he sweeps down the coarse rushes. A family of six once kept near one mower for more than a week, watching the marsh as he cut the crop, for at times he was able to kill a mouse or two, and thoughtfully laid them on a heap of stuff: whence they were sure to come and take them. And when the hot July sun had sunk to rest and the mists began to rise from the river, they sought some heap of stuff near by, where they roosted for the night; indeed, half-loads of litter are a favourite roosting-place, as are lonely posts later in the season, when the litter is all poled. And so on, from season to season, the busy kestrel kills off the vermin and works for the farmer, who is often thoughtless enough to shoot him: indeed, this hawk should be carefully protected.

CHAPTER LXII

THE OSPREY

It was a bright May morning on the Broads, warm, with a gentle breeze rippling the still glassy mere, which was a beautiful mosaic of colour, the blue waters being pied with light pea-green patches of colour, sub-aqueous fields of weedy lamb's-tail, where, in the spawning season, bream and roach already had laid their eggs. The rudd were rising, snatching the flies hovering above the blue wavelets, leaving ever-widening rings upon their ridges.

As I pushed out of a reed-bed, where I had spent the morning watching for otters, the bright scene entranced me, for though a familiar sight, it looked fresher than ever, and the breezes blew cool after the close air of the reed-jungle; but I suddenly stopped, for there, sitting on a post some hundred yards away, was a large bird having all the appearance of a great horned owl, with a cast of the eagle in its features. And at first I took it for an eagle, with its bright head and breast shining in the sun, as it sat gazing over the mere sleepily. But as I paddled near, it stretched itself lazily, as a man does after a good dinner, in a superb lingering way, stood on tip-toes, raised its long strong wings languidly, and threw itself heavily into the blue air with a few short screams. And when I saw the short tail, and the slow, low, buzzard-like flight, I knew him for an osprey—a bird that is often seen in spring, summer, and autumn passing through, journeying afar from the tree-covered crags where they nest. Indeed, I think many of these birds do not nest for the first few years of their lives, but remain in gay bachelorhood or curious maidenhood.

As he flew lazily over the marshes, he put up a pair of redlegs, who, mistaking him for a buzzard perhaps, attacked him right boldly, driving him in haste over their domain of marsh to a distant gate-post, where he alighted lazily and sat ruminating an hour by the clock *sans intermission*. After his dinner was thoroughly digested, he flew lazily back towards the mere, going to windward till he reached the bright expanse of water, when he turned, and, sweeping steadily down, hovered some forty yards above the water in the manner of a kestrel, suddenly darting down and gripping a fish in his talons, which he carried off to his post, whence I started him, driving him off in great dudgeon, for he flew excitedly, screaming wildly once or twice as he followed a silver stream away into the grey distance.

But what a shambles was the gate-post! The decaying old log was smeared with blood, entrails, and scales from top to bottom, and the marsh beneath was one stinking charnel - house of the remains of rudd—all were rudd: decaying heads, tails, scales, entrails, and clots of blood and flesh lying scattered in profusion—a sight to make one sick; and yet there was the osprey's gory dining-room. In truth, two fresh fish lay there with their heads partly eaten off. And then I saw why he likes our broads, for the rudd are plentiful there; and since they love the fly, and rise freely in warm weather to take them, they would, from being so close to the surface and intent upon their dinner, form an easier prey for his dinner; and this he knew right well, so he never caught any but bronze, gold-mailed, red-tinged rudd, and they are food fit only for him and his kind; but the fish-eating bird should improve his table manners.

CHAPTER LXIII

THE CORMORANT

OR "Cormorel," as the fenmen call this cross-tempered grotesque-bodied bird, pays occasional visits to the Broadland. You may see a herd of six or seven flying high from the sea on a grey winter day, their flight away across the sere marshes, resembling a bunch of geese; but as they draw near the cold glistening broad, where the stars are reflected, you see their wing-beats are quicker, and recognise the "cormorels," who have come to take in fresh water, and perhaps a supply of fresh fish.

But whenever they come in from the sea they seem restless, moving from one broad to another, and even separating, some solitary birds going a-fishing in the dikes, the banks of which are spotted with coltsfoot, and wherein the fish are filling with spawn.

But rarely do they come in parties: perhaps it is because they are so quarrelsome; for if, on a bright day, you see a party a-fishing, and two try to alight at once on one of the posts marking the channel, they will begin to fight fiercely with their powerful bills until one retires humiliated. More generally they are seen alone.

Mayhap some fine spring morning, as you sail across a broad, gliding before the wind over the liquid planes, the watchful cormorant, on a distant beacon, quietly looking at the growing sail, shakes his wings, stretches his neck languidly, and once more plays the sentinel; but as the white sail comes over the water, he rises on his legs, throws up his stiff tail contemptuously, and shoots a large dash of whitewash-like dung adown the front of the post as he flies lazily off to a distant post, secure from guns and other devilish machines.   Such is the reception given to the gunner by the cormorant.   But should the Broadsman get to windward of him and maim him, he will, if captured alive, seize upon his person, either hand or thigh, it matters not which.

But he and his kind, the shags, are best seen by the rocky shores of Mona.   There all winter you shall see them sitting on the rocks at low water fishing.   Uncanny birds they look, until their fierce-looking heads betray their tribe. When the grey mists veil the hills of Wales, and Bangor lights peep through the mists, often of an evening, too, you shall see them flighting there down the Straits to the island coast, flying black and swiftly against the azure, in bunches of eight, nine, or ten.   Or, on some bright sunny morning, you may watch an old cormorant sitting on the Bell-light surveying the Straits, a sentinel to the passing fish.   But he is essentially a bird of the sea, and therefore not a resident of our swampy lands, and so not much is known about him: he smells too—*que voulez vous de plus?*

CORMORANTS' NESTING-PLACE (*Chislebury Bay*).

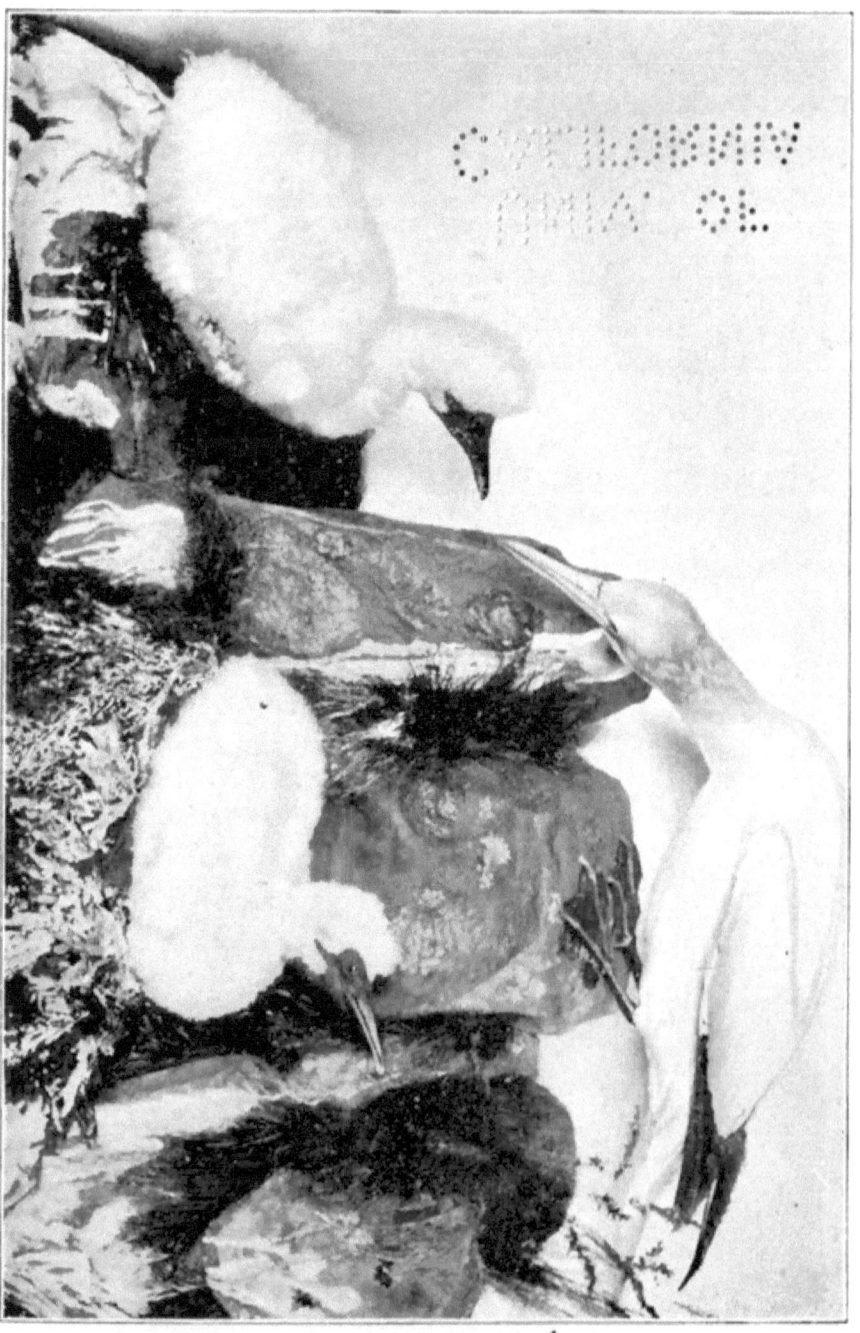

CHAPTER LXIV

"GANTS"

IN the grey autumn mornings, as the North Sea luggers
are hauling their nets, filled with blue and silvery herring,
from the cold water, amid the wild *laka-laka-lakas* of the
gulls, the *craa-craa-craas* of the gannet will be conspicuous—
as are the birds themselves—and a welcome augury they
are to the North Sea fishermen, who read in their presence "herring about here!" The fishermen tell of two
"gants"—the black and white gant—probably the immature and adult birds; and though they welcome them,
still, when the nets are being drawn in, and the *craaing*
gants go in a body, some hundred yards off, to lift and
strip a full net with their powerful bills, robbing the silvery
booty, they are not altogether pleased, and smile with satisfaction as some incautious youngster gets entangled in the
lint, and is dragged aboard drowned. And when his catch
runs to five or six lasts of fish, there will be hundreds
of "gants" and gulls robbing him of his well-earned
harvest. And perhaps it is this habit of picking up the
nets, and "deeving inter 'em," that has suggested as a
revenge the not uncommon practice, when they hang idly
over the fishing-boats, of nailing a fish to a board, and
casting it upon the troubled water, when the "gant," who is
flying a hundred yards above, closes his wings, and plunges
down, striking the board with such force that either his bill
transfixes the board, and he is a grotesque prisoner, or,
more commonly, his neck is broken from the blow. And the
fishermen laugh a laugh, which is a sweet revenge. But the

mocking laugh is answered by the *laka-laka-laka* of the gulls, and the *craa-craa-craa* of the " gant's " relations, whose cries will fill the grey welkin when the dark sea-stained boat again begins to haul in her nets, now drifting with the rushing tide, upon which the greedy "gant" and gull rest from their labours, sleeping as calmly as we would upon their down.

CHAPTER LXV

THE HERON

AN old cock heron stood alone on the silent snow-field beneath a grey sky, the setting sun burnishing his fierce warrior head—sable plumed, and flashed from his silver-grey back, spangled with sable epaulettes, that were lost in his pied and plumed cuirass. Patches of milk-white melting into the snow suggested his neck and thighs, and dark lines outlined his sharp dagger-beak and stilt-legs; for the heron is a true son of the Fens.

Not another bird was to be seen on that wintry eventide. Frank, however, as the fenmen call him, was not alone. A rising and falling patch of colour over the snow-field had attracted his attention. A hare was leaping across the white fields, making a dining journey to a planting a mile away. After gazing sharply at the hare till it reached an opening in a hedgerow on the upland, Frank's long neck doubled and he drew himself together. He was cold, for he and his kind feel the cold keenly, and yet they linger with us, faithful to the marshland.

As the blackthorn bursts into bloom, Frank goes off to his fir-trees by Reedham (for there is a heronry there, though few persons know it), or, more careless, builds in the low willows by the river, and he is rarer on the marsh-lands until August, when he comes with his young to the broads. Just before harvest you may hear the heron all night calling hoarsely, " Frank, Frank ; " also the shriller cries of the young as they hunt for eels in the dikes or by the mere-side, scarce a foot deep, leaving their dung on the soft warm grass beds, and showing the fenman where to set his springes; for Frank cannot conceal his traces—he is bound to leave dung or crush the succulent gladen growing in the yellow-lilied shallows, so betraying his fishing haunt. At such a season, and in winter too, you may sometimes see twenty or thirty together fishing for eels, eating them as gulls eat herring; and I have seen one throw up five large eels after he was shot, three of them being still alive.

If you wish to watch Frank's family fishing, you must "hide up" to leeward behind the reeds, and you will see them, their bodies nearly in the water, looking down the dike with one eye; but if one rise all will follow. Go just at shutting-in time, when the partridge is calling and the snipe scaping, the rails snoring, and the goat-sucker has left his day perch, for that time and the early morning, when the marshes are white with mist, are his favourite meal-times; but upon a moonlit night, when the dikes gleam like silver, and the broad is a polished mirror, Frank will fish all night long, darting his dagger-like bill in and out of the soft black silt, and striking his prey on the ground a few fierce pecks, then throwing it up and catching it head first in his long sharp-boned beak.

From Michaelmas to Christmas, when they are tamest and most numerous, you may see them flight from the hover, or car, where they have dozed away the day, to their feeding-grounds, flying from five to thirty yards above the ground, calling "Frank, Frank," as they near their favourite dike, where they alight suspiciously, and have their supper ere they flight back to their roosting-place on some dry hover, or in a planting.

Though Frank is dangerous when wounded, the fen-men say he is not so vicious as a bittern, though more powerful.

Like the fenmen, Frank may be seen all day in autumn about the marshes, walking along the dikes with his neck "reined" out, peering into the depths for eels; but he is only to be got there when "onsighted" by the wary gunner, for sometimes the fowler is sharper than Frank.

But sometimes he escapes with a broken wing, and if caught young, is made a pet of. I once saw such a bird living in a little garden with a toy-terrier on the most intimate terms. Flora the terrier teased Frank by barking at him, but Frank gave reproof with a few playful pecks of his sharp bill, and Flo' was silent. The gunner who captured young Frank had shot his mother; but he took the body home, and for days used to put dead eels in her beak for his captive, who took them from the dead mother, but would eat no other. Indeed, Frank's sole occupation seemed to be eating, birds' entrails being a favourite dish—even dead mice being acceptable.

One day Frank escaped, and a labourer tried to drive him home; but Frank was ungrateful and flew at him, at last finding his way home himself. But, like the "dear gazelle," Frank finally died and was mourned, for he was a good sportsman.

The heron is a graceful bird when on the Fenland; but let him alight on a tree, he is the most awkward bird imaginable, and looks as if in constant dread of falling off. On the other hand, he looks grand when fighting bravely high up in the sky against a head-wind, his neck set well back and his long legs stretched behind him, as his great sails flap to and fro against the blasts. But sometimes he is beaten; and I once saw one resigned to his fate, being carried like a balloon before a hurricane out to sea towards the great waste of the German Ocean, to arrive on the Continent wearied and dazed.

The heron is the decorative bird of the fens, and all through that delightful district you may see on all sides natural decorative panels worthy of Hokusai—one of the most beautiful being an old cock heron flighting against a pale rose full-moon across a reed-bed of amber and purple, calling desolately to the swallows, who have just gone straight up into the skies in a purple column, for they were leaving for the south.

As the red wintry sun fades from the glimmering landscape, Frank shakes himself, cries hoarsely, "Frank, Frank," and rises lazily into the misty air, his long legs stretched out like stilts, as he flies a few yards above the snow to a planting, now fading on the landscape, to roost. But no mate answers him, for many of his friends are lying frozen beneath the stars, and others had been feeding by a wake on the glassy mere, in company with coots and rooks.

Next morning, at sunrise, Frank comes back to his feeding-place, flying some twenty yards above the greenish lighted snow-fields ere he alights, with a hoarse, melancholy "Frank," by a little spring in a frozen dike, the rustling yellow reed-stalks screening him from the cutting winds.

I saw one walk along such a bank, with neck "reined" out in hopes that some one had broken a hole in the ice and the frost had turned up some dead eels; but he was disappointed, for he turned and settled himself down to watch at his station by the spring.

Soon another living thing appeared black upon the snow, creeping up stealthily behind the thin and rustling reed-screen. The black mass stopped, there was a muffled report, a puff of smoke cleared away slowly, and Frank lay dying. I could see his arched-neck with bristling feathers drawn back ready to strike at the eyes of the retriever, who bounded forward to seize him; but at a sign from his master, the dog stopped, wagging his tail, and the man approached the bird, and struck it a deadly blow with the stock of his gun, muttering, "You old hog."

"Why a hog?" I called.

The keen sharp face of the gunner turned slowly, as he began to reload his old bess, an ancient muzzle-loader.

"Why a hog?" he drawled, in sing-song, as he measured Frank's wing-spread—6 feet 2 inches. "Why a hog? Bless his old flesh, old Frank Linfort stole the muck-fork, and he'll eat anything."

He had poured the powder down his long gun-barrel, and was driving home the first wad, when he came on, and, looking at me, said—

"I know 'em, sir, better than most folk. I hev kept several tame ones—young birds they were. One I had ate suet, a wiper nigh twenty inches long; while I had a young one ate a roach weighing fifteen ounces, eels all you like up to half a pound, frogs, rats, mice, stanacles, and dead birds. They like their own sort best; they are death on birds. I do a bit of stuffing, and as sure as I get gutting a bird, in come my nabs and steal some. They're a bird wonderful quick of hearing, wonderful quick-sighted, and wonderful quick to digest their food—that runs through them like water." He paused as he poured his shot into his rough musket, but resumed—

"Old Frank ha' done me out of many an eel—the warmint—but I ha' cleaned his clock now, and I shall get tree bob for him. I only wish he was that white 'un what I be after one day last summer, nigh up to arter harvest; but he was too quick for pill-garlic."

And he picked up his prey, and stalked off in the snow upon a hare's track.

He had touched upon the curious traditionary opprobrium that Frank bears in the fen district, though I could never discover what muck-fork Mr. Frank Linfort did steal; but Frank is reckoned capable of divers dirty tricks, as the rooks and starlings know, for they may be often seen chasing him away from the sheep, but I have never found him guilty.

As the marshman said, "Frank will eat anything when pressed except white herring, and if it be indigestible, he will cast it up." Fishes' tails and feathers are often cast up in egg-like balls resembling exactly the castings of the owl. Many a one I have found upon the marsh-walls, the favourite dining-tables of the marsh-birds. As the fenman says, too, he is a good pet. I knew a young bird in captivity who would follow his master round when he took the gun, and behave just like a retriever, except that Frank ate the starlings shot for him. I saw him eat a pound roach and numbers of eels, which he killed by pecking them, holding them crosswise, then swallowing them head first. Upon one occasion he essayed swallowing two eels at once, but that was too much for him.

CHAPTER LXVI

THE BITTERN

THE " Buttle " is now almost unknown in the Broadlands, except in the hardest weather, when a few are shot; but a few years ago, a few pairs were to be seen about in the spring; and one spring night I heard the famous bumping noise, of which old gunners speak—a loud, booming, resonant sound, as of some one striking a brazen shield from afar—a haunting voice of the marshland. But altogether the "buttle" is rare, though four or five are shot every winter in the marshlands near the sea. Some old gunners aver that he nests there to-day, and a man once worked for me who caught a young bird alive in the sedge whilst egging in May 1864. He says, "I saw him in among the sedge with an eel in his mouth, so I darted arter him and caught him, and he was rare wicious. He'd a jabbed my eyes out if he'd a got at 'em. I took him home, and got eleven shilling for him."

One old gunner, who has found their nests and shot many of them, gives this account of them—the old man is alive now, and quite trustworthy :—

"Buttles—yes. They fly like a harnsee—only a little quicker, more like the kitties, but head and neck pulled well back—feed like a harnsee. They live on eels and fish, frogs, and such like. They sleep on the hovers round the broads, and you see them mostly of a night and morning at flighting times, flying right low if there be any wind. If you put 'em up onest, and don't kill 'em, and then try and put 'em up again, you'll have ter go arter 'em afore they'll go up.

"They'll light on trees, jest like a harnsee. I recollect one December mornin' I went shooting, that had been blowing and snowing. I had my traps in a bag, along with my old bess ; and up went a buttle to the top of a tree in a planting, and I shot him. Lor', you would hear a lot of 'em calling *gaw-gaw-gaw*, in Catfield Fen when I was a boy! I remember shooting tree in five minutes there onest. They are wonderful good eating—eat like a pheasant. I recollect onest one came to meet me with all his bristles set up and eyes open, and I knocked him over with a gun. They're wonderful tame things—not shy like a harnsee— a dorg will take 'em, but he must look out for his eyes. They allust go straight for your eyes. They live chiefly on eels, but they go walking about the land hunting for worrams. You can allust tell the hen bud, she be larger nor the cock. They are wonderful fond of sharp frosty nights. They only bump in spring-time. I never heard more than four bumps at a time—three loud bumps, and the fourth soft. That is done by putting their heads inter water.

"That were many years ago : I found their eggs. There was a reed bush in Catfield fens, and I went inter it and found tree buttles' nests there."

He has never found any since, but thinks a good many have nested in the Broadland, but cannot be found. The little bittern I have only seen in "glassen boxes," as the Broadsmen contemptuously call the "set-up" specimens.

COMMON BITTERN.

CHAPTER LXVII

THE STORK, SPOONBILL, AND SHELDUCK

THE stork is a legend of the Broadland to most men, for on all sides old gunners and marshmen tell you such things as follow :—

" I seed one many year ago a-setting on rush-hills."

And another :—

"One set on one of the posts in the broad, and Olfred went ter get his gun, but that was gone afore he come back."

And a third :—

" One was about the marshes here for a long time ; people seed him feeding on pike, and they went to Mr. Booth, and he come and shot him."

And Mr. Booth's book corroborates this statement.

The last I heard of as seen in the Broadlands was in the spring of 1889, but never a stork have I seen. I can only say there is plenty of evidence that at rare intervals they are seen about the district, but to most they are as legendary as the phœnix.

THE SPOONBILL.

A rare visitor passing through the Broadland is the spoonbill. The last I heard of was on Breydon in 1892.

But most old gunners will tell you with sparkling eyes of the few red-letter days in their lives when they shot a spoonbill; and one week I heard of five being shot out of seven who were feeding on a dike at East Somerton, one Shrovetide some six or seven years ago. Nor was it strange, explained the old keeper who told me, " seeing they live on the fresh-water shrimps and stanacles, and the dikes are full of them, we know."

The Common Shelduck.

I know an inn where a ruddy shelduck stands in a "glassen box," and a gunner tells me he shot one three years ago on the marshes near the sea. He says, too, he always sees them singly, and generally in the month of May. Old men tell how they used to breed in the warrens, but my own observation does not go beyond the bird in the "glassen box."

CHAPTER LXVIII

WILD GEESE

IN the hard winter of 1890–91, when the rivers and broads were frozen for weeks, I saw my first wild geese coming over the cold white land, with their tame-goose-like voices —cheerful as a band of good-hearted rovers driven from their country—flying along in the conventional wedge as they came off the sea, but quickly changing their figure as they spread over the Broadland and alight on the grass marshes—vast " fields "—their select feeding-places. Pink-footed geese are the commonest frequenters of the Broadlands, fifty or sixty at times being seen on a marsh, feeding over the lowlands, or in very severe weather upon the open fields or uplands.

And when these geese arrive the gunners are alert, and the news go round that geese are about, for they may keep for weeks and even months about the marshes in a hard winter.

And the gunners get into their grey punts the colour of water, scull up the dikes edged with ice, and get a shot at the huddling flock, for they often sit quite close, and will even let a stalker walk close up to them, being tame, until you get within shooting distance ; and then the heads that have all been up gazing at you rise higher, and the leader sounds his warning note, and away they go, forming a V, as do many wild-fowl when flying. But the V formation is of bad augury, as the gunner knows full well, for it bodes a high flight, near the stars, and generally a foreign journey across the grey sea.

But the wary gunner, as I have said, sculls up in his punt, and sometimes gets a shot over the wall, bagging a fine nine-pound bird or two ; or else he waits till the flighting hour, when the stars begin to peer, and, standing all ready, he listens keenly, and hears them coming, calling just like a tame goose, and suddenly the large birds fly almost silently above him (as far as their wing-beats go), when a red flame darts across the night, and some large bird falls into the water.   Indeed, many gunners say if they didn't keep calling they would often get past unobserved, though they only fly some twenty or thirty yards above the reeds at this hour.

And so roaming restlessly (like all wild-fowl) from marsh to marsh, night after night, they get knocked off by twos and threes, until often a single bird—the sole survivor of a flock —is to be seen, by day perchance, until he too is bagged.

That other goose that sails by the Broadlands, and rarely alights on the marshes, is what they call the "polean," black or Scotch goose, or, in other words, the brent, whose flight resembles a cormorant.   Indeed, the one or two I saw on the broads in the depth of winter we first took to be cormorants. A few have been shot on Breydon, feeding on the duck-grass.

Besides these two geese, bean geese and grey lag geese are said to be shot ; and Mr. Everett, of Oulton, tells me he has seen the bernacle goose there.

But the pink-footed goose must be looked upon as *the goose* of the marshland.

WHITE-FRONTED GEESE.   (*From life.*)

CHAPTER LXIX

SWANS

" O'er what vast lakes that stretch superbly dead,
  Till lashed to life by storm-clouds, have they flown ?
  In what wild lands, in laggard flight have led
  Their aërial career unseen, unknown,
  Till now with twilight come their cries in lonely monotone !"

E. P. JOHNSON, *White Wampum.*

THE swan travels with a false character. He is reputed by poets and other romancers to sing when he dies, and to marry respectably and live as "pure" a life as a respectable British grocer. He does neither. He dies with a hiss, and is as little of a monogamist as the Grand Seigneur. In February, according to swan chronology, the unpaired mate with each other; for already have the family fights taken place in frigid January, and the young have been separated from the old—the young, lone and desolate, having taken up their grounds ere they begin to fight fiercely and pair in February.

The 22nd of February 1891 was a lovely, silvery day, with a gentle breeze blowing from the S.S.W.—a breath of amorous spring that made the lusty mavises pour forth their full hearts. Even the hedge-sparrows and tomtits joined in the pæan to spring. A gentle breeze scarce ruffled the still waters of Buxton millpond on that sweet morning as a young pair floated languorously upon the silver water —for though they do not lay the first year, they tread— the young cock bird looking at times up the bright river towards the old church, upon the reflection of which an older pair of birds were swimming leisurely.

Anon, seized with desire, the old cock of the church pool paddled majestically, with all sail set, down to the mill pool, never deigning to glance at me as he passed. On he went straight to the young pair of lovers, and singling out the hen, he paddled with arched-back neck round her, and in a clever manner separated her from her simple lover, who sailed swiftly away. Turning his prey, he chased her before him, she paddling dog fashion, her heart beating with fear. Having placed some fifty boats' lengths between her and her young but timid husband, he rose on the water, mad with desire, and flew noisily towards her, and in some forty yards came up with her, seized her with his strong bill between her white shoulders, she, terror-stricken, flapping her wings and paddling towards the quaint low water-side, where she succeeded in landing, but never for a moment did he relinquish his grip. Meanwhile her outraged husband, who had come up with great show of fury and bustling feathers, got out on the bank beside them, and only watched and waited till the old cock writhed off backwards, when he stretched his neck, hissed, shook himself, and walking down to the river, began throwing water over himself, after which he swam leisurely off to his lawful spouse. The injured husband made a pretence of following the ravisher to avenge his honour, but soon desisted, and the injured pair began bathing and flapping their wings. The young cock was in all probability too poor in condition to give battle; and the end of it: the ravisher's lawful spouse received him joyfully.

One hour and twenty minutes afterwards the rape was repeated within a yard of my boat; he had grown bolder. So much for the swan's virtue and everlasting constancy.

That same day I saw the young couple "ringing the bell," as the fenmen say of ducks. Bobbing their long necks up and down in courtship, with all the grace of courtiers, a proof that she was uninjured; for if a swan be maimed but a pin's point it is taboo; that is an example of their flunkeyism.

It is said they do not pair nor lay until the third year. However, when they do begin, it is early in the year; and in February you may see the old cock-bird building those mysterious cock's nests, either to encourage the hen to begin, or else to serve as armchairs for himself whilst she sits upon the piled-up gladen, warming the large pound eggs into life, an operation which he shares dutifully. Year after year they choose the same locality for their nests, on a hover or on a rond close by the water, choosing the handiest material for its structure, generally old gladen or water-rushes; indeed one spring I watched an old cock build two "cock's nests," one of rush, the other of gladen, on hovers beside the water, after which both began building the green rush nest on a lovely little islet girt with a fringe of tall water-rushes and " cocks'-heads." With their strong mandibles they soon completed the floor of the nest, round which they placed a coil of green rushes, passing the building material to each other with their bills. And I had leisure enough to study the structure, for the birds leave the nest and go far abroad to feed until the first large egg is laid.

But as soon as the first rough-shelled greenish-tinted egg is laid they keep near at hand, though they take the precaution to cover the egg. Whenever at that season we approached the nest, both the gigantic birds clambered on to the nest with stupid hissings.

Every morning (for, like most wild birds, the eggs are laid in the early morning) we visited the nest, and noted that the eggs were slow to accumulate. Sometimes three days passed between the laying of the eggs. 'Tis the same with the French and English partridge. When the seven eggs were laid, the first proved, as usual, the smallest of the batch, though it weighed exactly a pound, and measured nine and a half inches in circumference, and was nearly five inches long, and green with digested subaqueous lime and vegetable matter upon which they feed. If robbed discreetly, they will lay as many as twelve eggs. The fenmen, who rob their nests and eat their eggs, steal upon them in the half-light, and quietly push the hissing, flapping birds into the water with their quants, quieting them with a smart blow of the yew-fir pole given smartly across the skull. But even then they will follow his shadow, hissing, and making great pretence of attack, for they are veritable swashbucklers. Indeed, their fights seem to be settled, as a rule, on sight, the younger birds fleeing before the elder, who merely drive them away if they invade their station or broad.

But none but a fenman would rob a swan's nest, for a nastier morsel than a swan's egg I never tasted—a most nauseous thing. After each egg is laid, you may see them "ringing the bell," and going round and round each other in circles, now widening, now contracting, their necks half bent, till they begin lining, with feathers all fluffy. I have seen them court thus strangely for over an hour, and when their love was consummated, they became all at once like white india-rubber birds, and began bathing, shaking their tails, and tossing the warm water over their glistening feathers, and drinking from the surface of the mere. After all the eggs are laid they begin sitting by turn, and in a month the young cygnets are hatched and taken to the water, leaving their piled-up cradle naked to the skies.

At such seasons you may hear the snorting calls of the old birds, and the replies of the cygnets as they feed upon the shallows over the hard bottom, for the swan dearly loves a "hard" covered with weed. There, too, in the lily-flowered pulk-hole, you may, if you chase the mother, see her sink her body into the water, and take her young on her back, sailing off in trepidation, whilst the cock comes with a flourish towards you. But never mind him, unless you fall into the water, when he would master you surely. I know one old cock who killed a big retriever easily in the water, and a man would fare no better if he had nothing to defend himself. For the cock will fight for his young—he is very fond of the fluffy birds—and you may see both of them pulling hair-weed and gladen roots from the bottom and feeding them. The young, too, catch flies on the water and eat them; and both eat fish spawn—there is no question of that, though the old birds will never take fish from hooks, like the great crested grebe.

At the end of a year these young will be able to fly, when two years old they will be white with swansdown, and at three they commence to breed.

In hard winters they are hard set in the wakes, and are often fed by keepers. For though many are only marked by having their webs punched with a wadding-cutter or slit with a knife, they are considered private unless they can fly, when the fenmen count them as fair game; and many a shock-headed fenman sleeps soundly upon swansdown, and at times feasts sumptuously upon cygnet—a dish, I confess, to me no better than poor, ill-fed venison, which it resembles in flavour.

The mute swan is the largest of the family, and next comes the whooper, and lastly Bewick's swan.

The whooper, with his wild trumpetings, comes but rarely to the Norfolk coast, and only in such weather as leaves him a mere bag of feathers and bones; but the small swan of Bewick is common—wedge-shaped flocks of five to ten coming across the frigid seas in a sharp winter, and dropping with a *yap-yap-yap* from high in the sky into the sluggish waters of the Broads—one gunner in 1890 killed

ten Bewicks, three whoopers, and five mutes.   One year
he killed six Bewicks at one "shoot" of the mighty swivel,
but could only carry five home, leaving the sixth on the
ice until the next day.

When flighting over inland waters, these birds rarely form
the V figure—that is reserved for long journeys across the
seas.   And you shall always know the flight of a wild
swan, for his wings do not *creak*.   The tame swan that flies
betrays his presence even on the darkest winter night by his
creaking wings—no matter how accomplished an aëronaut.

From afar in the winter evenings you may hear them
approaching, calling to each other like a flock of tame
geese, and did they but keep silence, they might pass
the lurking gunner on their arrival from the cold North.
But when they stay about the Broad district in hard
weather, they fly low, just above the reeds, at the flighting hour — often singly, and often in little bunches—
seeking for pastures new ; for they love a grass marsh,
eating the succulent crops as eagerly as tame geese.   But
on the approach of spring, never a whooper, never a Bewick
is to be seen.   They return to their homes, to tell of the
desolate lands and inhospitable inhabitants, who receive
them day by day with burning tongues of flame and deadly
rains of lead, and only the "collector" is happy.

Strange white birds, that grace our rivers and lakes—
now sailing like full-rigged ships over the smooth face of
the waters—now asleep in the bright moonlit misty night,
with heads lying on wing ; at another season sitting closely
upon some rushy islet like marble statues in the wilderness
of reed and water, or again penned up in a frigid lake in the
crystal ice.   Embodiment of graceful motion, lover of cooling
waters, farewell!   And mayest thou "evolve" that dying
song—often sung about, never sung—thou lovely flunkey
—for thy presence is thy be all and end all—thou peacock
of the waters.

CHAPTER LXX

THE MALLARD

A DUCK out of water is the embodiment of vulgarity, recalling a fat, big-boned, coarse woman, and the drake recalls a variegated specimen of the horsey fraternity; for he is loud, has a certain stop-and-look-at-me air, and cannot walk. In the water, again, these birds have a mechanical-bird look, and hence the power of decoys; whilst flying through the azure they look merely grotesque, as though they had no business to be on wings. And yet when, in the cold winter, you hear their precise wing-beats in the long nights, and their hoarse cries reach you from the dark reed, you realise what is melancholy. Such a feeling as the falling of the rain in a plantation gives you : a lonely sense of coming woe and desolation.

But a fen-bred flapper is the juiciest and sweetest of table-birds, for all your migrants smack of the sea-grass, a flavour no sauce will kill. The home bird is bigger, too.

In my opinion, the wild duck mates for life, though they are reported to pair in February, as I think rooks do, and many another bird who is reported to pair for a season only ; for you may see them in pairs all the year through, after the rearing of the young is done and the young flapper can do for himself; for in the education of his progeny the mallard takes no part, the husbands going in noisy bands to enjoy themselves, whilst their spouses rear the family on the marshes and in the reed-beds. Nor do the hens show much originality or ingenuity in selecting their cradles, for they often make the rushy down-lined nest on the top

of a river-wall, in a clump of rush, or by a broad-edge, only
taking pains to hide their eggs when they go off to feed
by putting some mixed rush and feathers over them. And
like many monogamous birds, the male often commits adul-
tery and rape in captivity, as well as in the wild state.
I have observed this fact; and I have seen a gander rape
a duck, as well as a swan rape another's spouse. Year
after year, if not disturbed, they will return to the old
spot to nest, generally in the heart of a rush marsh
near a dike or broad. Often the ice-bound rivers still
ring to the skater's steel when they begin to lay, and
fenmen have found their eggs "hard-setting-on" before
the middle of March. Sharp frosts and destructive rainfalls
do not prevent them laying nor sitting either. When the
hen-bird is sitting in meditation—her head drawn down into
her shoulders, and all but her tail and back buried in soft
down—the drake is not far off, often feeding on weed in a
neighbouring dike; so cautious is she, that you may walk
within a yard of her, and though she can smell you, she
will not budge; but if you do put her off, she will feign
lameness, or play some character in adversity. And she acts
with reason; for ducks, like rabbits and hares, give no scent
so long as they remain still. But even acts of evasion are at
times worthless; for sometimes the drake betrays the nest by
hovering over it, signalling danger to her. But, as a rule,
should you wish to find a mallard's nest, ignore the antics and
calls of the drake, and watch the hen-bird, if you can get a
sight of her, and when she drops into a marsh, look well
over the moist rushy knolls; or should she alight on a wall,
search the dry cars of bramble, sallow, and alder; or should
you disturb her on her nest, never expect to find it at your
feet, for she is wise, and when danger is near slips off her
hidden eggs and runs fifty or a hundred yards before she
rises on the wing, quacking to tell you where the nest *ought*
to be. She has been known to nest in low willows, and take
her young down in her bill when the day arrives for the

family to seek the water; and that is very early, for the
young ducks, who break their prison walls themselves, will
run out of the nest directly they are hatched, if disturbed,
though sometimes they will pretend not to see the intruder. Before the exodus to the water, the hen carries
off the old shells. Sometimes the drake assists in this
exodus to the water; he and his wife directing the young,
walking ahead through rush and gladen, across lonely dikes
and over lonely walls to the reed-beds growing in the shallows. But after that he is off with other husbands on the
moult, and the wife does the education down there in the
water, teaching them to catch flies and midges. But they
never swim far from the duck, the twelve or thirteen fluffy
balls keeping close to her, and diving like stones if startled.
Tame drakes will kill ducklings hatched off under hens;
this infatuated tendency, perhaps, drives him a-roving to
prevent crime. However, he evidently doesn't love his
offspring.

Wild duck are persistent layers; if robbed early in the
season, they will lay again even unto two clutches.

When the broad is white with lilies in July, and the
young have grown nearly as large as their mothers, they
attempt to fly, and once more the drake condescends to
return to his family; for the fenmen, with their long muzzle-
loaders, will soon be on their track.

As soon as the young flappers are strong on the wing,
at the end of July, the morning and evening flighting begins,
the family flying from one yard to one hundred above
the marshes, according to the weather, flying high in clear
weather and low on muggy days, where they feed and sleep
by day, and returning to the broads at nightfall, when you
may hear them feeding in the reeds, and especially on
moonlit nights, when they shriek and quack and enjoy
themselves immensely. The wily trapper knows the hours
between seven and nine at night to be the best for his
work. Now that the flappers have grown stronger on

the wing, you notice how they always rise head to wind, rising straight out of the water by sheer strength, and flying head to wind till well up, when they can throw themselves down to leeward very quickly on a breeze.

As the nights " pull in," and the gunners grow more keen, they begin their morning flightings to the sea, where they spend their days floating on the German Ocean, secure from punt-guns and "old besses," only returning to the land at night. Sometimes a fine day in September tempts them to spend the day on the broad; but the expert gunner knows well in such cases they will grow restless about ten o'clock in the morning, and make a bold flight from one broad to another, and he lies in wait for them, neatly "hidden up" in the reed. Again, if the weather be very coarse, they do not go to sea, but keep moving about the inland waters all day, for rough weather unsettles them. That is the gunner's chance; but, after all, it is not so propitious as the fine dawn of an Indian summer day, when they sit deep in the tranquil waters dozing, whilst the astute gunner sculls up to them silently, the rising autumn sun directly behind him, the hinder part of his punt being kept end on with the sun, and, ere they know it, a rain of deadly hail and a thunderclap startles them into shrieks from their dream; for most wild-fowl avoid looking at the sun.

At such times you may see them flighting in family parties; but the gunner soon breaks these up, when they flock in large parties, some old drake leading them. And it is useless to try and tame them. Should you wing any young flappers, the young, if captured, will nearly always drown themselves by diving when approached. I knew of some confined in a wire-netted horsepool : every bird committed suicide.

Some of the less affectionate families break up sooner than others, and you may in August see a single flapper swimming about in the warm dike water; and I've come across them wandering about the outside-marshes, and swimming through the lazy dike-weeds.

After the equinox's roar, and the cold gales blow from the nor'-western hank, large flocks of migrants come over and swell the ranks of the native mallard for a time ; but the home-bred birds generally, in their turn, leave the district to the emigrants. If the wind has blown fiercely for days, and it suddenly drops in a night, you may see every expert gunner out betimes the next morning, before the sun melts the crystals on the dead thistle-crowns. But he must not be there too early ; for if he fire into the birds' nightly feeding-places, they will desert, and seek new grounds and new lines of flight ; for they are peculiar, and won't be shot at at table, whereas by day they don't seem to mind the sport. No; you must wait till they congregate in the early dawning on the misty water ere they flight seaward—that is the time to fire at them with the loud-echoing swivel-gun.

And on a winter's day you may sometimes catch families going seaward all day—herds that have come from inland waters to the westward.

But when the broads are laid and the land is white, the ducks leave altogether, going south with their warning note of coming cold and desolation. But even at such times a few lazy birds remain, but they are scarce worth powder, being merely bone and skin. In sooth, they are true augurs of frost, increasing in numbers just before a great frost or thaw.

Often in winter, when the frosts have broken, you may in the early morning see the silent gunner working up to fowl on the still water; keeping the dark sky to westward, he will paddle to within twenty yards of a bunch of mallard, having passed silently through bunches of pochards and coots, the gunner always keeping the dark sky behind him; for on a winter's dawn, before sunrise, the east is clearer than the western sky as seven is to one.

Let us watch yonder long cigar-shaped boat sculled noise-lessly through those coots rocking on the water; for there is a good breeze, and it will be a difficult task for the gunner to-day to bag yon mallards rocking up and down close together in a bunch by yon reed-bed. So far the chances are for him; for in still weather they sit scattered over the water. He draws up noiselessly; you suddenly hear a tap on his boat-bottom, his iron toe-plate, and the birds rise frightened, heads to wind, the old drake calling loudly, signalling the company. Still the patient and experienced gunner does not fire, but, raising his swivel-gun, keeps them covered as they come together, and when they have risen some thirty yards from the water, a bright flash lights up the alder cars, a loud rumble, mingled with quacking and squawking, followed by the sharper reports of the shoulder-gun, settles the business, and you see mallard lie dead on the water.

The gunner is a patient man and full of restraint; for though he saw a feeding marsh as he came along that morning, the feathers, pulled grass, and trails betraying one of their favourite tables—a flooded marsh bottom—still he fired not at them; he only flushed the birds, but came on to meet them on the water, which he has done to some purpose.

But behold he is pushing to the hover, and in the gloom you see him, like a ghost, peering into the tall, silent, dead reed-stalks. In a twinkling his gun is to his shoulder, and the report echoes over the misty waters, and by the rising winter's sun you see him push into the reed and recover a wounded bird, now dead. And when he shows it you, he points out where the creature has plucked feathers from her breast and stanched her breast-wound, as is their practice; and as he tosses her into his well-filled punt he mutters, "They allus do that, they allus leave the water when hit. 'Pears as though they was afraid of bleeding to death in the water."

And such is the case; for if not too badly wounded, mallard "tarn fish" and dive, and swim under water to "onsight" the gunner; and they invariably make for the shore, swimming with their bodies just under water, only half of their heads showing on their outstretched necks, until they reach a resting-place, preferably on the dry ground. Sometimes they recover—thanks to their rough surgery—and at times you may come across their relics or their tracks on a dike, where they have been pulling duck-weed—a food not so dear to them as is generally supposed. Indeed, they prefer grass, worms, snails, and corn of any kind—either wheat, barley, oats, or beans. After the oat-harvest on a moist marsh stubble, you may find their moultings and feathers in the dampest furrows. And if the mallard is awkward and vulgar-looking on land, yet is he good food, especially the home-bred flapper.

COMMON DUCK ON NEST. (*From life.*)

CHAPTER LXXI

THE SHOVELLER AND PINTAIL DUCKS

A FEW pairs of "shovel-bills" are to be seen all the year round in the Broad district, but never many. Rarely I have seen three, four, and even six birds together, but as a rule they hunt in pairs.

The shovellers are shy birds. At a distance they look, to the inexperienced, like mallard, but they are swifter on the wing, and their necks longer and thinner. At the flighting hour, on dark grey winter days, you see them fly over a broad or down a river-course, looking as dull and grey as the landscape; for the cock does not get his elegant plumage till the spring. One spring I watched a pair for weeks—the old cock floating on the edge of a small broad— for the cock, as a rule, swims against the stuff. When flushed, the hen would invariably start off first, calling, and followed by the handsome drake, both flying toward the marshes by the sea. But they left the district before they nested. Indeed, I have never seen a shoveller's nest in Norfolk, though old Broadsmen tell me they have taken their eggs.

THE PINTAIL.

That capital table-bird, the pintail duck, is rare in the Broadlands, but once seen you will never forget him. His long neck and tail are conspicuous as he flies singly against the low line of the winter sky, but you will rarely see him; indeed, old gunners, who have followed the fowl night

and day for years, see very few, and never more than four or five together, sitting on the winter waters with the "smee," birds that are said to look larger than the pintails, and "ter act more busy, for they allust take ter the wing first."

PINTAIL DUCKS.

CHAPTER LXXII

TWO TEAL

TEAL dislike the open water, and love straggly reed-beds and stuff growing near plantations abutting on the Broads. They are the shyest of birds, especially if they have once been near a decoy-pipe, starting up when frightened, the leader whistling—which shows them to be wide-awake and on the alert—the others juggling, something after the manner of a snipe. As they grow more accustomed to the situation, the leader's whistle becomes shorter and less frequent, finally dying away to a long soft whistle as they re-alight in the stuff. If feeding in large numbers, as they do in winter on the less secluded broads, and you once again disturb them, they fly up in parties, the leader of each party whistling in alarm as they fly round and round their feeding-place. And should you not disappear, they will finally fly high into the air, and disappear swiftly in the grey atmosphere ; for the teal is a very swift flier, as the flighters know, for, as they say, "the teal come past you fit to take your head off." And unless the flight-shooter has light enough to discern them coming, as they fly low and swiftly, he seldom has time to pre- pare for them, and they go whizzing past like cannon-shot, leaving the gunner swearing.

When the Broads are laid and the marshes snow-fields, the teal go away, and do not reappear till the frost has gone. But they are plentiful in winter, when the water is open—most, if not all, of these winter teal being migrants, coming across the sea on a nor'-west wind, like most other wild-fowl. They come in large flocks. One gunner tells me he once saw a flock of hundreds come out of the sea like starlings. He shot across the flock with his old muzzle-loader, and killed ten birds. Though loving the stuff near the coverts, occasionally they venture in winter to the open water, spreading as they alight; but they are always rest-less in the open, rising and alighting again three or four times in as many minutes; quacking like a duck, only quicker, if it be daybreak; later in the day, they generally whistle if disturbed. The knowing gunner will flush them in the open and wait quietly till they alight again; for they generally alight closer together after being flushed, and so afford a better shot. After that dread report the cripples take a bee-line for the shore, working out their lives with beating wings through the icy water, their bodies covered and only half their heads showing above the lagoon; indeed, the passage of the wounded birds through the water is more fish-like than fowl-like—their wings act as fins; and such is the practice of all water-fowl. In the spring many go away, and the Broadsmen say the "home-bred uns come back;" for they aver the home-bred ducks and teal do not winter here.

Howbeit in May the residents nest on the ronds, marshes by the sides of dikes, or in cars near the water, in similar localities to those chosen by the mallard. The nest, too, re-sembles the wild duck's, but is smaller, being made of down and moss and grass. When leaving their nests they hide their eggs, as does the mallard, for warmth, and to hide them from the quick eyes of rook, harrier, or buzzard, no doubt; for nine eggs form a valuable possession, and nine teals' nests out of ten will have nine eggs in them, though eight and ten are occasionally found. When incubating, they sit very close, and will allow you to tread upon them at times. They are tamer, too, at this season, allowing you to ap-proach them if feeding in the dikes—squatting rather than flying up, as is their custom in winter.

In some parts of the district these nests are rare, but

in others they are not infrequent, though never common. Should you find a nest with the old bird sitting and put her off, she will feign lameness to lead you away from her treasures.

When the young are hatched, the old bird takes them down to the dikes or broad, and educates them in the reed-beds edging the lagoons, teaching them to eat duck-weed and other food such as the mallard feeds upon.

After shooting comes in, the clutches get scattered; and in September you may see them singly or in pairs, rarely in full clutch, frequenting the gladen and reed, the alder cars, or slads upon the marshes; but by Christmas either the home-bred birds have gone or they have gathered into bunches. Yet at this season a pair may be flushed (for I believe that, like the mallard, they pair for life) from a dripping alder car, or else a lone widow or widower may jump up from a clump of stuff; but the ubiquitous gunner does not give the lone one much chance, for its body is sought after, though, as a table-bird, I think the teal to be greatly overrated. Even the plumpest home-bred teal of the year cannot bear comparison to a home-bred flapper—mallard or widgeon. I am not "wrapped up" with them at all for eating; but tastes differ.

GARGANEY TEAL.

The gay "gargle-teal," as the Broadsmen call this con-spicuous bird, is first to be seen in the month of April, when you may also hear that peculiar sucking note of his. Some fine day in that month, as you push round the fresh young gladen, where the big lily-leaves are spreading over the water, you may flush a pair of gargles, the cock-bird glistening in the sunshine. Away they fly—the drake then calling—towards the marshlands, for they have come to nest; but now-a-days the few pairs that come over are rarely allowed to hatch off. The egger and collector, whether poacher or keeper, is immediately on their track. And since these men know their breeding habits to be exactly those of the mallard, his nest is the more easily found. The situation of the nest varies a little; for the gargle prefers an inside grass-marsh or a wheat-marsh, preferring some quiet spot by a water-dike, where she lays her nine or ten eggs, and hatches off her young brood, which resemble ordinary teal, taking them down to the ronds and reeds by the water's edge as soon as they are hatched, decoying them as does a mallard. Nor will you be likely to catch sight of the youngsters until they are strong on the wing, for they rarely venture out of the shadowy palisades of reed before they have attained the dignity of "full flappers." But when that day shall have arrived, you may see the clutch feeding on the brackish broads, and the gunners about sometimes shoot eight or nine out of the ten or twelve that constitute the family.

And when the herring-fleet begins to put to sea, the few remaining gargles go away again, full of reminiscences of eggers and gunners; and the students of birds are left with memories of rare visions of birds or nest.

NESTING-PLACE OF TEAL AND SHOVELLER.

CHAPTER LXXIII

WIDGEON

A STRONG September nor'-easter brings the smee over in considerable flocks, which scatter, on arriving, over the dikes, ronds, slads, and broads, living singly, in pairs, or bunches. Nor are they shy on first arriving, but allow the insidious punt-gunner to approach closely. But his terrific reports soon frighten the birds, and they grow shy and begin to use their tongues, the leader whistling his soft warning, when danger threatens his party. Should a widgeon be alone, however, or in the company of gulls, or with perhaps two or three of his own tribe, he grows anxious, and at the least approach of danger stands up with head erect and calls in short, quick whistles, for he is now excited and nervous. Experienced gunners tell me they can tell roughly the number of widgeon present by the leader's whistles. But with all his whistlings and anxiety, the flock is often shot whilst asleep—an easy death, at any rate.

After having grown shy they seldom alight in a bunch on the lagoons, but straggle all over the cold grey waters. In the dark they resemble mallard, both in their flight and wing-beat; but they are generally known to the flight-shooter as they approach him by their leader's whistling—a whistle resembling the babies' toy rubber-ball with whistle. On a dusky winter's night the flight-shooters are at times kept busy, as they come over the flat-land and reed-beds in pairs or fours, and even bunches of one or two score; for they flight every morning and night from one broad to the other, and sometimes during the day—that is, if disturbed when they happen to be resting.

Though they have their particular feeding-places, they, like mallard, go away in hard weather, returning when the frost has gone. They are, however, more restless than mallard, and seem to require a greater variety of food, to judge by the frequency with which they change their feeding-grounds, though, when resting on the water, they behave just like mallard.

When wounded they make for the rond, just as do other members of the family, and, like them, they are their own surgeons, stripping the down from round their wounds, or pulling out their wing-quills from around any wing wound.

When the kingcups open and the spring is here, they often repair, I am told, to the grass marshes, and eat grass just like a goose, thus growing fat. Nor are they in any hurry to leave us, for I have often known them stop till May; and old gunners tell me they have at times seen one, two, and three about all summer-time. To my certain knowledge, a widgeon was shot in this district as late as the latter end of June.

For the table I prefer a widgeon to a teal or mallard— unless it be a home-bred flapper stuffed with potato or celery, roasted whilst fresh in a brisk oven, and served unspoiled by sauces. Such a dish is only fit for hardy sportsmen, and they are about the only people who get wild-fowl served in this fashion. Your *chef* keeps wild-fowl too long, and then ruins them with sauces.

CHAPTER LXXIV

"POKER" DUCKS

THE Norfolk gunners love the "poker"—he is their great mainstay in the hard winters; and capital eating is the "sondy-headed poker," almost as good as a canvas-back from Chesapeake Bay.

In the language of the Broad gunners, there are five pokers:—

1. Sandy-headed poker (common pochard).
2. Red-headed poker (red-crested pochard).
3. Black-headed poker (tufted duck).
4. Scaup poker (scaup).
5. Golden-eyed poker (golden-eye).

The red-crested pochard may be dismissed at once as very rare. One gunner I know has only shot one in his whole career.

The golden-eyed poker, too, is rare, though far more common than the red-crested poker. He comes over late with the cold winter, and the flighters know him by his very swift flight and the peculiar rattling of his wings as he flies. These birds herd together on the water in small parties of nine or ten, and are very shy and difficult to approach; so that the gunners pick their day for shooting golden-eyes, preferring a fine still morning, when the sun rises clear in the winter sky. After the sun has risen for some half-hour, the crafty old gunner launches his long punt-gun, and keeping the bright sun right behind him, he approaches the herd, and the loud reverberations of his long gun tell of the death of some golden-eyes, which fetch good prices, and that is all he cares for.

Commoner is the scaup poker, a tame duck at all times, that arrives in large flocks in September, and later on in driblets of ten or a dozen. If the bunches of scaups on the water be small, the gunner is able to scull up in the grey dawn as near to them as he likes—close enough, indeed, to see the whites above their broad bills; and when they rise, just as do the other pokers, he pours his deadly fire into them, bringing the heavy ducks down with loud splashes into the icy water, whence they are gathered and sent to market. And so from September to March (when they leave) the scaups are occasionally killed by the fowler.

Commoner still is the tufted duck or black-headed poker, or black and white poker, which arrives on the Broads at the end of September or beginning of October, flighting to the feeding-grounds of an evening, just in the same way as the common pochard. In winter they seem to be shyer than in the early spring, for the punt-gunner often cannot get within shot of them; but in early spring I have rowed quietly up to within a boat's length of them, as they swam by a reed-bush, where they feed as does the common pochard. Their flight, too, resembles that of the common pochard; but they are a far more beautiful bird, their coats gleaming like velvet and their bodies shining a greenish-black on a sunny day, as they rise by flying along the water, gradually rising till their legs clear the surface, when they rise up into the air, and turning, go before the wind " like thunder," as the natives say. At this season, too, they will sit on the broad all day, merely flying up as a boat passes, wheeling round and round till it has gone, when they return to dream, an the noisy coots and waterhens will let them. But they leave us in March, though I have been told that they have bred in this district in recent years.

And last and most common is the sandy-headed pochard. When the October gales begin to blow, the "sondy-headed pokers" begin to come over the sea in large or small flocks, as the case may be, frequenting the icy waters of the lagoons by night, feeding on their favourite hair-weed, and leaving again at daybreak, though some few erratic females come to feed by day. Like most wild-fowl, when they first come over they are not shy; but those streaks of fire across the dark skies, those roaring thunderbolts, and the shrieks of their dying companions, soon warn them the beast, man, is about, and they learn caution.

At nightfall the flight-shooter lurks in his boat, screened behind a reed-bush, listening intently for the roar of their wings; for mallards make a *wisp, wisp*, with their wings, but pokers and tufted-duck come on like the roar of the sea—and well it is they can be heard from afar. A mile off they betray themselves; for the gunner has merely to cock, and stand with his gun to the shoulder, waiting till the birds are upon him; for they fly at dusk, though if frightened, will not flight till it is dark. And then, if he fire, they will fly high, beyond the reach of his deadly shot; but if stormy and windy, they fly low, and if he be alert he may get a few; but they are the most difficult of all wild-fowl to shoot —swift, and strong, and short-necked birds that they are.

After they have passed these lurking flight-shooters, when they come to the dully-glowing broad, they throw themselves down into the water, and make for the reed-bushes, where the water is shallow, in search of their dainty hair-weed. The knowing but reckless gunner knows this habit of theirs, and if unwise, will, on a still clear night, paddle to the easternmost side of the broad, keeping the birds between him and the clearer westernmost sky, so getting a good shot; but he does not do this often—even if unwise enough to try it—for if shot at whilst feeding, they will forsake the feeding-ground, and go elsewhere. Knowing this, the wise gunner waits till the winter's dawn, when they collect together, just before quitting the feeding-ground—that is his chosen time, for he knows they will not desert his waters, though they may take to leaving the feeding-ground earlier the next morning. A good plan, too, is to sail down to

windward of them, and they'll rise "right face and eyes
at you." Pochards generally sit well together on the water,
but are best to shoot at when they first rise from the
lagoon—that is, if you are going before the wind yourself
—and gunners knowing this, wait till they turn off; for
they do not rise like mallard, widgeon, and teal, who jump
straight from the water, head to wind. On the other hand,
pochards of all kinds cannot rise directly out of the water,
but flutter, run on the water, and flying up, turn and go
before the wind, the female often calling *auk, auk, auk*.
Again, if they sit in straggly groups on the lagoon, the gunner
waits till they rise, for then they come together, and he
catches them when they are twenty yards up, drowning the
roar of their wings with the roar of his mighty swivel. The
cripples make straight for the shore, as do other fowl, only
they always keep by the water, whereas mallard, widgeon, and
teal will stray far inland over the walls into the marshes.

The gunner loves the "poker" because he does not desert
the frozen lagoons, but, with the coots, frequents the straggling
wakes. Nor do they desert the Broadland till everything is
frozen hard—then they too go, for they must eat. And the
gunner kills numbers of them in hard weather; for if shot
on the wakes, they merely fly up, going round and round
over the frozen broad, but soon returning; but if the gunner
harass them too much, they'll go off in search of a quieter
wake, and if such be found, they never return to the dangerous
wake first frequented. They are curious, too, and will come
right up to a wily gunner hidden in the stuff, coming in one
or two at a time, and when he taps on the boat, they fly out,
and he gets a shot. Shooters, too, say you can in the dawn
scull up to them, and by putting your hand over the side of
the punt you can sometimes catch them, for they will come
to it out of curiosity; whereas a mallard wouldn't come near
you, for he can smell you.

On bright moonlit frosty nights hardy gunners go down
to an unfrequented broad, where there is a wake or rain-

channel open, and place themselves facing the cold clear moon, firing at the pokers as they go across the bright disc. I have known one gunner shoot thirteen in this way.

The majority of pochards go away in March, but I am convinced from reliable testimony that the common pochard has bred more than once in the Broad district. An old fenman solemnly swore to me he saw seven pochard's eggs taken from Sutton Fens; and further, he declares they were brought off under a hen, but all the youngsters died. A gentleman assures me he saw a common pochard duck and her brood in one of his covers near the Broads ; and lastly, a gunner, whom I have never known to lie, and who is thoroughly reliable, tells me he has shot them every month in the year—and that last August (two years ago), on the 1st, he shot a hen-bird. I have myself seen a hen with internal eggs as late as April 21st.

As a table-bird, the poker is highly valued by some ; but I prefer a mallard, though the pochard tastes more like the somewhat overrated canvas-back duck. I think a good fat home-bred flapper mallard comes before either. But they are all good.

COMMON POCHARDS. (*From life.*)

CHAPTER LXXV

THE GOOSANDER, MERGANSER, AND SMEW

WHEN the waters are clear and cold, when the reeds are yellow and their tassels frayed, when the cold breath of winter is upon the face of the land, and the wild-fowl come with melancholy cries from their frozen homes, a few goosanders arrive, and several mergansers, in flocks of seven or eight, for the merganser is the commoner bird in the Broadland.

The goosander, indeed, rarely appears except in the depth of winter, and then he is generally alone, and a shy bird to boot, diving on the least alarm, for he is an expert diver and fish-eater.

The merganser (called locally the "sawyer") is, however, more sociable, though a voracious fish-eater, and as cunning as clever. But these birds are comparatively rare visitors, and rarely seen except by professional gunners, and those real lovers of the Broads who woo their mistress in the depth of winter, and even they are only favoured with glimpses, as the birds dive and rise again in alarm.

The smew I have never seen, for he is perhaps the rarest of the three; but one, two, and three have been killed in the Broadland during recent years. One gunner I know shot one four years ago, and another gunner killed one two years ago, and upon another occasion during the same winter he wounded two more out of a party of seven. But the wounded birds escaped to some low ronds, and could not be found at all, though diligent search was made by the sharp-eyed punter.

CHAPTER LXXVI

WILD PIGEONS

THREE kinds of pigeons are found in the Broadland, though one, the "blue rocker" or stock-dove, is rare—the two common kinds being the "ring-dow" (wood-pigeon) and turtle-dove.

THE RING-DOVE.

In the showery month of April the large flock of old ring-doves, who have been feasting upon the freshy drilled peas and beans, or frequenting the banks green with turnips, separate into flocks previous to the nesting; but even then they love the new-lays, and eat voraciously the young corn just sprouting; indeed, that is the place to get a shot at them. At this season you may see them flying swiftly about the covers, looking for a nesting-place mayhap; for about the middle of May or beginning of June, when the marshes are yellow with crowsfoot and the lokes gay with hawthorn, and the plantings full of their soft coo-rooings, they begin making their simple platform of twigs, placing it high in an old thorn-tree, or preferably a conifer, if such is to be found; but these trees are rare in the Broadland. And when the two pure white eggs are laid, you may, perchance, see them from the ground through the frail nest. And when the young are hatched and the woods resound with the cooings of the old birds, the farmer is brought under requisition, for his fields are ransacked for peas, beans, wheat, barley, and oats, and thus many fall victims to the watchful gunner; but if the field of cloth of golden crowsfoot has been thick and luxuriant, you will see the old ring-dove (for both take a turn at feeding their young) there gathering the crowsfoot seed, a dainty they are very fond of.

And should the ubiquitous urchin rob them, they will build again and again until the fourth time; indeed, I have seen young nestlings in the middle of September. But should the first young brood escape all the dangers of childhood, they will linger about the honeysuckle-decorated plantings long after they can fly; indeed, they do not leave the nest before they can fly, and are still fed by the old birds; for their education is slow, and they are fully grown before they seem able to feed themselves. And should you disturb them before leaving the nest, you will see them fly out and drop to the ground awkwardly. And when the yellow harvest is ricked they again gather in flocks, together with fresh arrivals from over the sea, and seek the new-lays, where flocks of a thousand scramble, and prove very destructive to the newly-sown seeds planted after the barley has been cut; and if scared up by the farmer, they fly to the nearest plantings, where they sit on the trees thickly as leaves. And if the day be foggy and grey, they will mope, sitting upon the bare branches, the watchman sitting on a low branch, altogether making a delightful harmony in grey, such as Mr. Whistler might love to paint. And should you draw nigh, you will see the motionless leader arise and shake his wings sharply against his side, his signal to the moping, dreaming flock that danger is nigh, and of one accord they arise and wheel away to some bare oak in the heart of the wood.

And at all times they are good eating, which is more than can be said for some of the farmer's enemies.

THE STOCK-DOVE,

miscalled the "blue rocker," is a smaller bird than the old "ring-dow," as he is swifter of flight; and as he is often to be seen in his company during the winter, the difference is noticeable. But in spring-time they separate into pairs, or may be seen flying about singly, especially on a windy day, for he seems, like the peewit, to love a windy day. They, too, are injurious to the farmer, dearly loving a freshly-set bean-field or ripe corn, and more dearly oats; indeed, in their thieving propensities they much resemble the old ring-dove.

But in nesting they choose the old rabbits' holes on the sand-hills, where on a fine day in the summer you may at times be startled to see a stock-dove walk out of a hole and fly away in search of food for his young. But they do not nest in the sand-holes so commonly as of yore. I think they prefer the less frequented dunes; and I remember finding several nests by some wild dunes overhanging a marsh yellow with ragwort growing in waste patches.

In former days, when they bred more commonly in Norfolk, the gunners loved to lay wait for them on the marshes, where they used to go in search of seeds for their young. And later, when the harvest-moon is big in the sky, they flock into the plantings of an evening with the old ring-doves, whose company they continue in for the rest of the winter, until the beautiful green-burnished patches on the marsh shine in the spring rains, when they separate, as I have said. They, too, are delicious for the table—far sweeter than many a bird of greater repute.

TURTLE-DOVE.

When the marshes are growing green, recovering in the genial April air from the scorching frosts of winter, you may in your walks come upon a flock of turtle-doves feeding leisurely in the bright sunshine—a shy flock, too, for as you approach they fly up, their white-tipped tail-feathers glittering in the sun, and you remember at once the turtle-dove's soft cooings, birds that have come to build, in June, their little stick platforms in the budding thorn-trees, their favourite nesting-places; or they choose the catkined sallows resounding

with bees, or, more rarely, upon some thick branch of honey-suckle, bright and sweet with trumpet-shaped " sucklings."

At such times the little plantings by the water resound with tender cooings, and the thin bright air is cleft as a single bird flashes past in search of seed for the young—those young that they feed with half-opened wings, ejecting the mashed food into their very crops 'mid the family hubbub.

And later, you may see the pair of youngsters sitting upon some turnip or mangold balk, and should you show up suddenly, they will run, and so swiftly that without a dog you could not catch them.

But 'tis little you will see of them at any time, for in September they are gone, and, unlike the ring-dove and stock-dove, they prefer a serener clime to winter in—a clime where they can fight; for, as every one knows, the dove is no angel, but a hot-tempered, hot-passioned bird, whose mate at times delights in arousing his ire by planting herself with her tail to the wall when he is most amorous.

But the doves are not Broadland birds, though they are not rare there; still, down in the marshland, we get glimpses of them, as they rob the marsh - farmer in large flocks assembled, or gather crowsfoot-seed upon the marshes, or fly to and fro from planting to planting carrying food for their young, or in the spring-time fly about the plantings —the marshland and the broadland is not their home, 'tis merely a district lying in their way, or else a place whence they may gather food—they are birds of the cover. And in such they delight, whilst their croonings, resounding far over the marshland and the water, adds to the charm of the district, be they heard in early spring, or in the after-glow of summer, when the warm autumn bursts of sun delude the more amorous into the belief that spring has come again. Voices from the upland are the voices of the doves to the dwellers in the marshlands, sweet afar-off wood-notes stealing across the great waters when the frail leaves quiver and are lost in the crooning of the summer sea.

Q

CHAPTER LXXVII

PALLAS SAND-GROUSE AND QUAILS

FENMEN tell you of flocks of these birds coming in from the sea at long intervals—flocks of thirty and forty flying in such manner that they think them to be plovers at first sight; but as they draw near, their flight and chatterings soon proclaim them to be sand-grouse. The last invasion was in the summer of 1888, when many birds fell to the share of the professional gunner. I only know of these invasions by hearsay, never even having had a meteoric glimpse of one in the open, such as one has had of so many other birds.

THE QUAIL

Has been rarely seen of late years, though old gunners tell me in former days they used to come across them, when they say they "used to get together like partridges, and whistle when they flew up."

I have never seen the quail in Norfolk, or, indeed, in England, in a wild state; but, having seen them commonly abroad, I can vouch for the truth of the statement that they get together like partridges—indeed their run is like a partridge, nor are they so shy.

But in the Broadland generally they are scarce indeed—mere passing visions, seen now and then by the ubiquitous gunner.

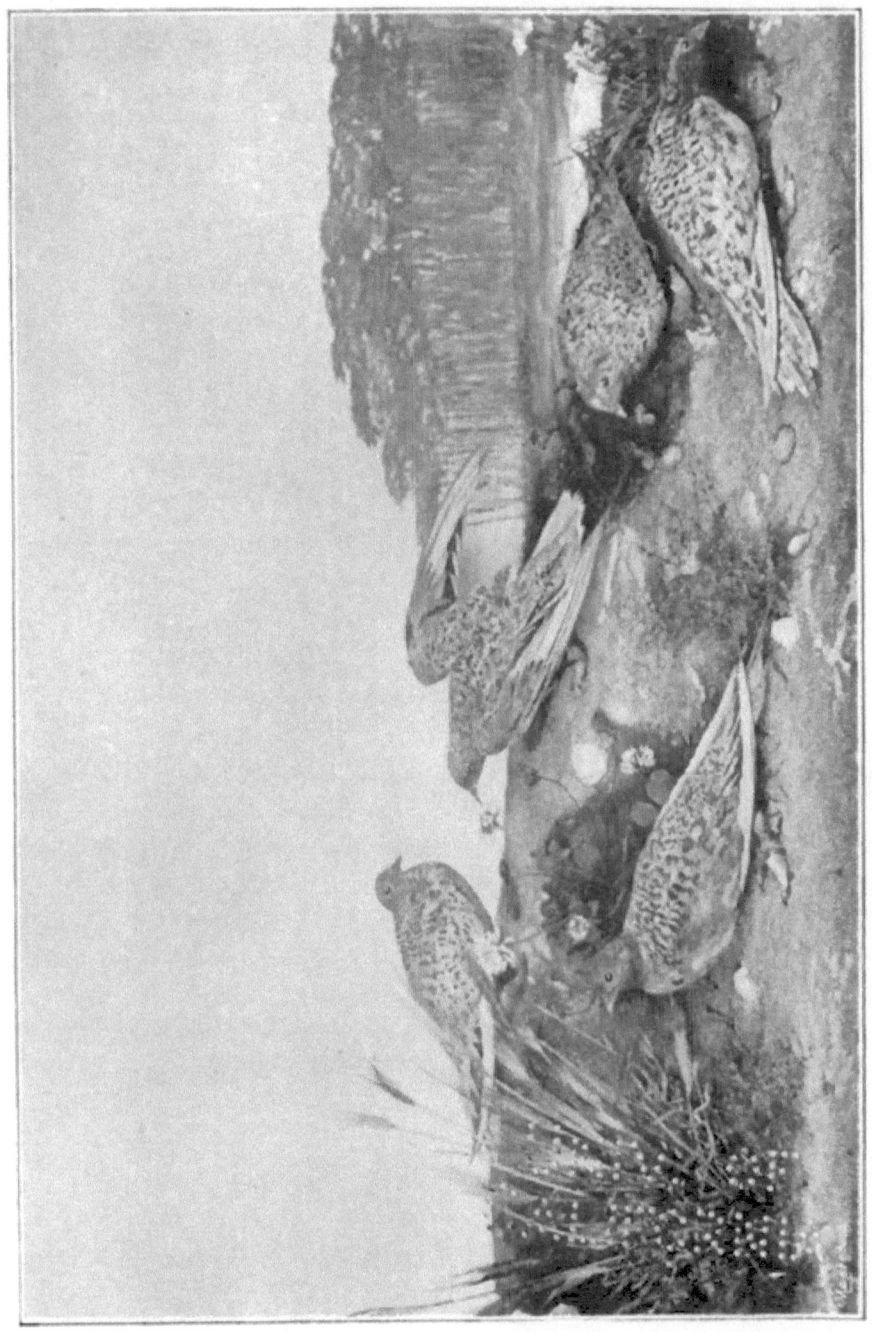

CHAPTER LXXVIII

THE PHEASANT

THE pheasant is a lover of the reed-bed and marshland, though this has been denied by the Cockney journalist. In winter, 'tis true, he prefers the warm cover by the water-side, but in spring he delights to nest on the open marsh, and to hunt for food through the reed-bed. When the early mists of spring wrap covert and lagoon in a grey cloud, you may hear the shrill proud challenge of the cock-birds, for they pair early, wandering over the marshes and along the wakes, and in the cars, till they find a suitable nesting-place, often laying on the open marsh in a clump of rushes, making their nest of leaves, grass, or rush on the ground, whereon they lay ten or twelve eggs. At this season—the middle of April—they lie very close, and you may tread upon them. The hen, too, when sitting, is not easily put off her nest. I have in fact lifted her off with a stick, she hissing like a snake all the time, and yet she would not leave the nest. Directly the young are out of the shell, they leave the nest and follow the old birds about the marshes, feeding on ants and ants' eggs, insects, and corn, when they can get it. The mother never leaves them till they are almost full-grown, when the clutch separates and scatters over marsh, reed-bush, and corn-field, wandering into turnip-fields and plantings in search of worms, of which they are very fond, and they even frequent hedgerows. But as winter draws near they prefer reed-beds, thorn bushes, and sallow bushes, whence they are to be flushed at the top of an autumn or winter day, as the acute gunners know well. On the marshes the breed is smaller, being the purer bred little marsh pheasant, which is the sweetest of all for the table. That the pheasant is naturally a lover of water is proved by his swimming like a duck if knocked into the water by the gunner.

Though the lordly bird, with his imperious voice, bright plumage, and bounce, is looked upon by the Cockney "lover of nature" as a park bird, he is a real marsh bird. The true English pheasant; the little brown pheasant, with the darker-coloured hens, is a marsh bird, loving boggy places, and often nesting in grassy tussocks surrounded by water.

CHAPTER LXXIX

PARTRIDGES

THE English partridge frequents the marshland in small numbers, though he is scarce, now that the marshes no longer grow corn.

Still a few are left, and in February you will see them in pairs, though they do not nest until the middle of April, when the hedgerows are green with newly-burnished leaves of delicate green.

At that season you may find dung and trails along the crests of the sandhills by the ever-sounding sea, a promenade beloved by the partridge, for there ants do abound. Down by the broad-edge, too, where a shady planting fringes the rush, is a favourite spot; for there, too, in the soft moist soil ants' eggs and grasshoppers are to be found in plenty.

Partridges, like ducks, hesitate not much in building a nest, a small hole scratched in the mould, sometimes roughly bottomed with dead grass, at others bare as your hand, suffices. Nor is she very choice in selecting a site for her rude cradle—a cosy corner of a long grassy balk; the top of a bank, thistle crested; a cool ditch by the white dusty road; a corn marsh; a grass marsh, where the land buntings lay; and at a pinch, the bare marsh beneath a tuft of bracken or stuff. In all these places I have seen the nest with the greenish eggs. Once I counted twenty-one eggs in one nest. Indeed, they will choose almost any spot with some cover overhead; for when flushed from the nest, she runs a yard or so, rises steadily on the wing, dungs, and flies in a straight line, away from

her nest, generally close along the cover.   But she is hard
to stir; for she sits very close, and is long-suffering.

In the marshes she loves to nest in the same place as he
loves to lie and sand-bathe. Any airy little place on the
marshes where there is a mass of beaten-down dead grass,
or a little heather growing, there her forms abound, and
dung shows her last abiding-place, eke if you do not flush
her and send her with scattering wings over the marshes.

And why should the nest be more elaborate, for as soon
as ever the young are out of the shell they leave the nest.
Indeed, an old gunner once saw one go forth with part of the
shell upon him, so eager was he to begin the battle of life.
Still, they are always alert to start and run and "hide up" in
any little car directly the mother gives the danger-signal.

And if you come upon the new-born clutch, the mother
will practise the arts of deceit, for she will fly off feigning
to be injured as to her wing. Sometimes she will do this
when flushed from her eggs, but more often she is too
scared. And when they begin their childish wanderings,
they seem at once to "do for themselves;" for though their
parents may point out their food—ants and their eggs,
insects, and corn—they do not feed them.   During this
period you may see her wandering on the oat-fields, and
later you may hear the old birds calling their young every
night; for until they get well-nigh full grown, the young
wander afield; but once they are full grown—and they
grow quickly when the corn is edible and the beans are
swelling, for they dearly love young beans—the family breaks
up, and scatters over the face of the land, to fall a prey
to the gunners, who do not always wait for the "first."
And as autumn skies close down over the face of the land,
and the landscape grows grey with fogs, the partridges
lose themselves, and stray into turnip-fields, into red-
berried hedgerows, or over the marram-hollows by the sea.
Everywhere they are to be met with, for in a fog they are
lost, and roam about solitary and desolate.

And at times they are struck by the gunner's shot and not killed, and though falling into the broad, they will swim like a duck for the reedy shore. And as the weather grows colder, they frequent the ronds and small cars dotted over the marsh, and when the pale winter sun shines feebly at the top of a winter's day, you are sure to find a partridge or two in the sallow-cars, thorn-bushes, or reed-beds, which the fowler knows well too, for the partridge is by no means a shy bird, though they do skulk in cars in the colder weather.

These birds always remind me of a peculiar kind of domestic fowl, so tame are they when full grown, so fond of corn and beans, as the farmer knows well.

As a table-bird, the fresh partridge is superior to a good young fowl, and to some wild birds—*e.g.*, the rails and plovers, perhaps; but since the disgusting practice of eating fowl in a decayed state obtains in this country, the partridge and pheasant are worshipped as table dainties just as is a lean tasteless piece of venison preferred to a good well-fed saddle of mutton.

The "Frenchman"

Is a born fighter; and though his English neighbour fighteth unto death, he is invariably beaten by his lordlier, fiercer brother. He is a more alert bird than the Englishman, not lying so well for the "sportsman"—altogether a shier, wilder creature.

Indeed, the gay Frenchman chases the English bird away from his dominion, and the English partridge does not thrive where the Frenchman lives.

The Frenchman is later in pairing, taking his spouse in the latter part of March, though he begins nesting about the same time as the Englishman; but he is a greater architect, choosing a hedgerow top, some ivy-crowned hedgerow bordering a lane leading to the uplands, often returning year after year to the same place.

Though a shyer bird, as a rule, than the English bird, in the spring the soft influence of love tames the Frenchman's heart, and he grows more hardy.

His hen does not lay her eggs every morning, but often slips a day or two between each effort; and when the young are hatched, the proud mother will fight for them, though she, too, is capable of making a ruse when hard pressed.

During childhood and youth they live the same life as the English birds; but directly they can fly they grow very shy indeed, often leaving a field by one gate as the sportsman enters by the other—that is, unless they squat, as they are prone to do on occasion. And when a " Frenchman " does squat, if you want to see him run past, lie down and watch him take to his heels, running down a furrow or turnip ridge like a racing bird, though he is really more of a marsh-bird than the Englishman; for he sits round about the marshes in summer by the edge of the land, and often builds in grass marshes, the hens getting their heads mown off for their pains, so closely do they sit upon their eggs.

They, too, can swim, though the young birds are sometimes drowned in the reeds where they alight, probably thinking it dry land, and getting their wings entangled therein as a penalty for bad judgment. The Englishman, however, is the better to eat, as everybody knows.

Gunners tell of a partridge much smaller than the Frenchman, though gay with his red legs. But I have never seen one.

CHAPTER LXXX

THE RAILS

THE land-rail is a rare bird in the Broadland; yet in May his voice may yet be heard in grass marshes—a voice like that of a bird-scarer's tool or Cheap Jack's rattle.

A skulking bird is the corncrake, as, indeed, is every rail; but if you see him within a few feet of you, he is bound to escape through the thick brown grass. Even in a turnip-field, which he sometimes frequents, he is hard to put up; he trusts to his feet in all but cornfields, where he seems to rise on the wing more frequently, as the gunners and harvesters know. In truth, the harvesters can catch him when the yellow swathes lie in golden bands athwart the marshes—for if he do not rise, he will run for the nearest swathe, and for once the men catch him and eat him afterwards in a dumpling or on toast, for he is capital, though overrated. Mowers in the haysel come across his nest, but I have never seen one; indeed the mowers, those chiefs of bird-nest finders, rarely come across them to-day, even in long-prolonged droughts, when they are commonest. When the harvest is stacked, however, the gunners have a better chance; and in September and October some are shot on the bare fields, and those that escape the deadly shot wisely go across the sea to far serener climes.

THE SPOTTED RAIL

Is the only name by which the spotted crake is known to the Broadsmen, for he is rarely seen to-day; indeed, the water-rail is far rarer to-day than the spotted rail was fifty years ago.

Old gunners tell me they have shot them in years gone before the end of March, and that they "allust used to come over then and on inter April, and that they never seed none in the depth of winter, as they have heard some gents talk on; they never shot none no later than October, that was their time for going away—September and October. They fare wise birds tew; they kinder throw their woices; but that's only done by tarning their head. I ha' watched 'em many a time. They allust used ter build in May, in the sedge or chate round the broads—they like the ground same as snipe do—they build a neast jest like a water-rail's, only a little smaller. Lor', I ha' found hundreds of their eggs in Catfield fen, and mostly allust they built their neasts of chate. They generally lay seven or eight eggs, but I never seed more than ten in a neast. They feed their young on insects and them totty little water-dodmans. But, lor', we used ter pay no regard ter 'em, only ter know 'em from a common rail; they be smaller, fly quicker, and are more humpy like, and both on 'em fly legs down. But the spotted rail is a quicker bird; he don't mardle about in the stuff, like the common rail. But onest, sir, I heard a strange creetur, a rail, what mewed like a cat. I told some chaps cutting gladen, and that got about, and, lor', lots of people used ter come and listen ter him. The mawthers thought it was a spirit; for the later that got of a night the louder that mewed. I found his neast, tew, on a grass piece." So rare are spotted rails to-day that I have never seen a nest, and have heard but few birds, but I have heard their peculiar *whëoo-whëoo* whistle, that will never be forgotten.

The first I heard was on Somerton Broad, some years ago. I was sitting, one hot August night, the breeze rustling my curtains, when I suddenly heard a loud shout, then a pause, and again came the whistle. I thought the old eel-babber, whom I knew to be near, was signalling; so I went out in the starlight and shouted—

"Hullo, there! What are you whistling for?"

"I wasn't whistling, sir," he replied softly; "that was a spotted rail."

And we both listened—the fish splashing, and the reeds rustling in the breeze, when again the whistle came from a gladen-bush near by.

Getting into my boat, I pushed over the sleeping waters, and through the dozing water-plants, when the whistle sounded on my left, farther away. For long I listened on that still summer night, hearing the bird first here and then there, now close by, now afar off, with its strange whistle. And the northern sky was growing bright with the breaking day ere I returned to my bed, that strange inhuman whistle ringing in my ears. I have heard their voices a few times since; but the bird is rare indeed to-day in the Broadlands, and almost extinct. Indeed, in another twenty-five years it too will be silent.

THE WATER-RAIL

Is a lover of moisture. No sooner does April come with its soft showers, bedewing the reeds and rushes by the Broad-edge, than you may hear the rails sharming—making that hollow, clucking, snipe-like noise that resembles water being poured from a half-empty bottle—a hollow, gurgling sound : the courting voice. If you be lucky, and get a view of the cock at this season, you will see him set his feathers out and gurgle in his glee. And if you wish to find his nest (and the end of April is the surest time for you to search), go you down to the ronds, or on the marshes, where the water lies an inch or two deep, and a fresh crop of chate, sedge, soft rushes, flourishes; or else seek clumps of broken-down gladen or amber-reed, or heaps of litter-washings fringed with foam, or laid rushes lying near the water, for in all those places the rails delight to build. And if you do hear a bird gurgle or groan, look well for their runs in the moist water-plants—distinct paths dotted with dung: paths that often lead you up to a clump of herbage where a nest is placed—but you may find several of these runs before you find the nest with the eggs, for the rail builds several nests before it lays. Should you fail to find his nest in this way, choose a showery morning or eventide, and go to the soft places I have told you of and listen quietly, and you will soon hear the bird—if there be one, and it have a nest—groan *boh* ! like a cow, this voice tells you that the nest is near at hand. Still listen, and you will hear the loud groan change to a less noisy twang (like that made by a broken banjo string), which is a sure sign you are "hot," as the children say; for his groaning and twanging is made just before treading, and only in the breeding season. Should you be lucky, and find the nest, made of reed-leaves or sedge, you may find any number of eggs up to twelve—but eight is the usual number—and you will see the nest is lightly covered in, the bird having pulled the stuff over as a roof, unless she have built it under a heap of washings or clumps of laid rush. And if you keep an eye on the nest from day to day, and you find the water rise, you will be surprised to find they have built their nests up —as does a waterhen—to keep the precious eggs above high-water mark.

Both birds sit, and are never far from the nest, the cock ever ready to groan and twang—even on moonlit nights—if anything disturb them. Directly the pretty young rails are hatched, they leave the nest, and move about the moist coppices of rush and sedge, feeding on the insects they pick from the water-plants or find floating on the water itself. But you will seldom see a whole clutch together; they scatter on the approach of danger and hide in the soft stuff.

They bring up two broods in a season; but, if robbed, will build five or six times, the new nest being each time built close to the robbed one. Nor do they delay about it; for, should you rob a rail to-day, in a fortnight you will find another nest close by with eggs in it, and you may find their eggs as early as Good Friday and as late as July.

When the water-grasses grow rosy, they leave their nesting-grounds, running about from place to place, often flighting some distance at night-time, as do moorhens—in truth, one can sometimes hear waterhens very high up in the sky of a night. But although they shift their grounds, they never go far from the water, and often betray themselves by their chicken-like squeals; for though they fly high when shifting grounds, they never fly much when in the stuff, but generally " run for it " when danger is about.

As the days draw in they begin to " perk " or jump up on to the reeds and gladen to roost, never flying up into low trees or bushes, however, as does the more cautious waterhen. And this precaution is necessary, for when the weather gets cold, and the stars sparkle with frosty light, the weasels, and rats, and stoats get keen set, and often devour them whilst asleep. But a stoat seems to fascinate them by day; for if a stoat come across them in a yellow reed-bed, the bird does not fly away, but runs on in a dazed sort of way, and is soon and easily caught by the hungry creature. Hawks, too, are fond of them in keen weather; and if they escape the buzzards, and harriers, and rooks in their childhood, the first hard winter that comes they fall an easy prey to other vermin. Gunners dislike them very much, because they warn other birds of danger, and some gunners aver they are friendly with the pheasant, to whom they always give the danger-tip. Nor do they behave differently when chased by men than when chased by vermin; for they first of all seek to squat or hide, so that they cannot be flushed without a dog, who often is able to catch them before they will betray themselves.

When the water—for they must have water—in the lagoons is turned to ice, and the stars sparkle, and cold winds roar across the flatland, they draw up to the drains, and sleep in holes in the river-walls ; but they are never safe anywhere, for the hungry crows, and buzzards, and hawks, and stoats find them out, and "trim them out properly," as the gunners say. Yet they are most friendly to men if undisturbed, and I know a painter who used to feed a water-rail regularly every morning with bread-crumbs, the confiding bird coming from the stuff on to the rond to feed amongst the robins, whom the bird after-wards chased away.

And he is good to eat himself—in a dumpling, or curried, he is a splendid fellow ; and it is a matter for regret that the hard winter of 1890–91 was the cause of frightful mortality amongst those birds—the vermin "cleared them up," and in the spring after that Arctic winter a rail's nest was as rare as a spotted-rail's had hitherto been. Never was such scarcity of rails known in any of their favourite haunts. I believe the water-rail will one day be as rare in the Broads district as is the spotted-rail to-day.

WATER-RAIL'S NEST AND YOUNG (*in situ*).

CHAPTER LXXXI

THE WATERHEN

THE waterhen is a fighter—the chief fighting water-bird; his very *crr-o-ook—crr-o-ook* is a challenge, his cry being wild and characteristic of the lagoons. In the middle of March, the chequered month, the cocks begin to fight for their mates, calling *crr-oo-k*, and nodding their heads at each other till their passion rises, when they fly at each other, lying back in the water and fighting with beak and sharp steel-like claws like game-cocks—such a splashing and commotion they make by the gladen and reed-beds—till their little differences are settled, when the victors go off with the hens, never having slain each other, however.

After they have paired, they begin to build their nests towards the end of April, and always on the water; for the waterhen, like the otter, loves to have the water near at hand to slip into, and, by diving, escape when an enemy approaches. These birds choose all sorts of places for their nests, but chiefly broken-down gladen and reed round the mere-side; more rarely they build in chate or soft rush on a wet marsh. The nest is a bulky structure, made of dead gladen, or reed, or sedge, or rush, according to the stuff surrounding it, and is so placed that, if the water rises, the bird can raise its nest —but how this piece of engineering is effected I know not, but do know that they raise their nests if the water rise too high. The most beautiful nest I ever saw was on a flooded marsh, placed in a tussock of sedge, the green sedge braided all around, forming a green bower exquisite in its dainty beauty—a bower hiding the eggs from the prying gaze of

man and buzzard, their chief enemies. This habit of theirs
of braiding over their nests with the young gladen and stuff
sometimes, however, betrays its position.

In the nest are laid the seven, eight, or nine well-known
eggs, when the birds begin to sit, the off-duty bird feeding in
the shallows near, diving and pulling up the weed from
the depths just like a swan. Later, you will see the bright
scarlet-crowned cock, and hen of soberer hue, swimming
in the stuff with their fluffy little nestlings—the old birds
whistling softly, and the young answering. On the least
alarm, the youngsters dive and swim to cover, where they
skulk, the old birds crying excitedly through the stuff,
warning the scattered clutch from the enemy's pursuit.
Though the young leave the nest soon after they are
hatched, they frequently return to it and roost, forcing
the nest down, and dunging all over it. But the prettiest
sight on still clear days is to see the little nestlings huddled
on a large water-lily leaf, or feeding across a mosaic of
leaves, old birds and all. I once saw a little nestling deserted by his family die on a lily-leaf. Should a buzzard
pounce down into the stuff and suck the eggs or take the
young, the hen will shriek dolefully, screaming round the
nest in the most agonised manner, her voice resembling that
of a human being in distress. I once caught a " buzzard " in
the act of sucking some waterhen's eggs.*

The young are difficult little things to catch in the stuff.
It is the practice with marshmen who wish to capture them
to mow a clear space round about their nest, and then
" hide up " near-by of an early morning, or at dusk when
they come out to feed, and pounce upon them in the
" clear." Yet I have known a well-trained dog carry two
in its mouth at once. I have known the same dog to bring
three eggs—one after the other—from a waterhen's nest
to his master, all unbroken. If you chase the young up the
dikes, and they get hard pressed, they will take to the rat-

* *Vide* " On English Lagoons."

holes for shelter. The old birds, too, often do this in winter if closely pressed, and are often caught by dogs in this way. Should there be no rat-holes by the narrow dike-way, the chased bird will suddenly disappear, and on searching the spot where you last saw him, you may see a few bubbles rise at short intervals; and if you keep as quiet as a mouse, you will see the artful creature lift his head above the water and look round as innocently as if he had expected to see nothing. Should he see you standing there, he will quickly disappear again below the placid water, and fly under water —indeed, in a clear dike, you may see him working his wings like paddles. When he thinks he has passed you, he will come up close in against the shore, just raising his head above the water, as he did previously, and looking carefully round him. If he thinks you are at a safe distance, he will stealthily glide out into the herbage on the shore and disappear. You see they have to be artful, for their parents leave them when they are half-grown, and they have many enemies. Even cats like young waterhens. I remember an old keeper, who lived amongst the stuff, told me he kept an old cat who for days kept watching a waterhen and her brood, till at last one day the curious waterhen and her chicks came out of the stuff and walked straight towards pussy, who made a light leap and captured a youngster, eating it. This happened day after day, till pussy devoured mother and all —at least so said the keeper.

Directly the old birds leave the first brood they commence building again, and hatch off a second family, and when that comes to pass their customs vary. Some families are happy, and the first brood of youngsters will come and help bring up their younger brothers and sisters— the two families and the parents may be seen swimming together—but in other cases the parents will none of the first brood. I remember one summer watching an old couple chase their fully-grown first-born across a broad whenever they swam out of the stuff—they kept the younger brood to the gladen-bush on the southern side of the broad, and chased their first-born back to the gladen-bush on the northern side of the water. And if the hen ever caught them feeding in the clear, as she often did—for they were fond of feeding on the lily-leaves and hills in the broad—she would peck and thrash them back to their gladen-bed. But, with all her care, her younger brood suffered. One day I heard a flop in the water, and saw a stoat swim a dike near-by, and scamper into the straggly reed-jungle on the rond—her preserves. Presently I heard her shriek, and there was such a commotion in the dry reed-stalks—the stoat was helping himself. When next I saw the brood, three youngsters had gone the way of all flesh. Nor are they devoid of courage, as I have said. I have known them, after being chased and maimed, turn on their backs, and spring at their captor's hands like game-cocks. They will do this sometimes before they can fly.

Like the rail, they build many "cock's" nests, and year after year they choose the same nesting-places, building and laying three or four times, if robbed ; but if not mulcted, they generally rest content with two broods. But they often have to build many times—for many of them are careless, building in the most exposed positions by water-dikes, thus often prolonging their family cares into the month of July. But they have to thank the rooks, crows, hawks, and "buzzards," as well as the eggers, for their unnecessary labours and cares. As the nights draw in, they begin to fly up on the reed and gladen and into low bushes to roost, out of the way of the vermin ; indeed, they will even fly up to roost into hawthorn-trees, and I have known of one nesting in a low thorn-bush by a broad. If the weather be cold and frosty, they go up the drains, and wander to the foot of the land in search of food and corn, which they dearly love. Should the frosts get severe and the broads freeze, they will, if not eaten by otters or stoats, hawks or harriers, rise up into the air and flight, like rails, to some more propitious feeding-ground. But always, when the weather is hard, they wander far afield in search of food and water, even going up to the marshmen's doors. In snowy weather they are often found dead or starving—a mere mass of bones and feathers, without sufficient strength to run or fly. But such weaklings generally fall an easy prey to vermin, who often track them in the snow. In snow, I have traced a stoat's trail across the marshes following the trail of a waterhen, and presently the trails ended, and a bunch of feathers and picked bones were the silent witnesses of the tragedy. The active stoat knows well in such weather they often have not strength to fly, nor indeed do they ever fly much at any time—a mode of motion oftenest practised when they are disturbed feeding outside the stuff, when they fly up, with drooping legs rippling the water as they flee into cover.

Beautiful as the waterhen is on the lilied lagoons, he is equally attractive in a dumpling. In a good curry, too, with nice dry rice, he is *sabroso*. Roast, also, he is capital; and though I have known sportsmen ask, " What shall we do with him now we have shot him ? " I prefer him as a table-bird to most other birds, rating him a good dish.

WATERHEN'S NEST AND EGGS. (*In situ.*)

CHAPTER LXXXII

THE COOT

I

IN the Broadland a smug, cunning, selfish, greedy man is contemptuously called "an old coot," and if it be said of a young couple that they go into reed-bushes searching for coots' eggs, their moral character is taken away. So the coot's character is not held in high esteem, neither is his flesh, as we shall see.

In March coots begin to draw into the reed and gladen beds for pairing; then there is a rare commotion in the coot world—callings and splashings, and runnings on the top of the water, and chases and fights with beak and claw after the manner of the waterhen, and rushings at each other with arched backs and lowered heads, and divings and flyings; a "noisy lot of varmin" they are in the pairing season, for they are ruder and more pugnacious than waterhens, those sleek, greenish-black, fat, strong, unpoetic birds, with their gleaming, heavy-looking head-plates and shining feathers. If you do not see them fighting and going on, you can hear them in the stuff, ere they come dashing out, the two cocks, mayhap, after a hen, one finally succeeding in driving her away and getting her to himself, whilst the conquered one goes back into the reed to try again. Towards the end of April they begin to pile up their swan-like nests of dead gladen or reed, choosing a sere gladen or reed-bed along the edge of the broad. The nest is a big one, and higher than the waterhen's, and seems to float on

the water, but, like the moorhen's, it can be raised, and is
raised if the water rise. In their nest is deposited from seven
to nine eggs, after which the birds begin to sit; and if you
are sharp, and go stealthily along in a punt, you may suddenly come upon the sitting bird, when she will "off nest"
and dive into the water like a waterhen. Whilst one sits
the other feeds round about near-by, diving to the bottom
and pulling up the weed, which it brings to the top and eats.
As soon as the pretty young coots are hatched the family
take to the water, and you may on a fine April morning
see them swimming about, just outside the reed or gladen
beds, but they generally keep inside the bush. But should
you come upon them suddenly, they will all dive and scatter
for cover; on the other hand, if they first see you, they will
noiselessly sink themselves under water, just leaving their
beaks and two eyes above. Sometimes a late snowstorm
overtakes the young brood, and they perish miserably. I
once saw two dead youngsters and six eggs half-buried in the
snow; the young were not strong enough to get away, and
the old hen wouldn't sit the storm out. As I have said,
they are coots and selfish. But when they have left the
nest they return there to rest at nightfall; it is warmer
and safer from vermin there. They are a fast-growing bird,
and soon after they leave the nest it is difficult to tell the
young from the old birds, though the old ones are greyer
and balder. When the young can feed themselves, the
parents leave them. During the breeding season, if any
unfortunate young ducks come in the way of the old coots,
they will wantonly kill every one, chasing them round
the reed and gladen beds till every duckling is a dead
bird; and for this pleasant habit of theirs they are often
killed off, especially round about decoy pipes, where
young decoy ducks frequent. Indeed, the coot seems to
hate the duck; and I know one keeper who robbed a
coot of her eggs, giving her some ducks' eggs to sit
upon, but she wouldn't sit upon them, and deserted the

nest.   They raise two broods in a year if left to themselves, but if robbed they will lay three or four clutches
of eggs, their eggs being found as late as July, when
there are often full-grown young birds of the year to be
seen.   These young birds hang about the reed and gladen
all summer, venturing beyond the stuff now and then; but
if disturbed, they go flapping, half-flying and running to
cover, whence their *ca-aks* may be heard on still nights for
a mile.

When the autumn gales begin to break down the dying
reed and gladen, and the frost to lay it, they venture farther
out into the open lagoon, flocking together—sometimes
six or seven hundred collecting together.   Experienced
gunners can tell them at a distance by their constant habit
of flying up into the air a little way and re-alighting on the
water.   And experienced gunners do not harass them, for
they seem to attract fowl to the lagoons, and themselves
seem to be attracted by any wildfowl, for directly a bunch
of fowl alights they go drawing towards them.   Should these
flocks be disturbed at one end of the broad, they'll fly over and
alight at the other end, so that they cannot be easily shot,
unless a number of guns go out, whose barrels they have to
fly across—that is, unless there be a strong breeze of wind.
Then they get blown towards the gunner, if he know where
to go; for they are weak on the wing for their size, and
cannot beat against the wind like other fowl.   The punt-
gunner gets a few shots at them as they rise, too, for they
are not strong enough to rise straight out of the water,
but must rise head to wind, and beat themselves up by
wings and feet before they get to the wind.   They rise from
the land badly, too; but in both cases, after they get fairly
on the wing, they turn down to leeward.   The waterhen is
stronger on the wing than the coot, as may be seen by his
frequent rising from stuff, and his rising straighter up than
a coot.

When the ice begins to form, they know whether it is

going to last, and form a gunner's augury. One old
gunner used to say, "There's going to be a long frost
soon—the coots will be off in tree or four days. I heerd
them give the call t'y mornin'." But they don't leave
the lagoons, as a rule, till they are frozen out—merely
huddling together on the ice, if not disturbed, till every
place is frozen clear and hard; for though they live on
the bottom hair-weeds, they will for a time put up with
the grass growing round the shores of the lagoons, if it
is to be had.

They are alert birds, and can smell danger quickly, often
frightening off wild-fowl by "splodding" about when they
see a gun-punt approaching. If chased by a hawk, they
dive, and throw up water with their feet as they dive
(though hawks never seem to strike at them in the water).
If hard pressed by man or dog, they may take to the rond;
but they are poor walkers, merely waddling along, balancing
themselves with their wings. At the flighting hours, in
freezing weather, they fly to and fro looking for water;
and if the gunner fire at them during these hours, they
are likely to leave the district. But some gunners do
not mind if they do leave, and get six or seven in this
way, themselves lying hidden up in the same place. They
roost on the gladen hovers, and take to the open water
before dawn, when they are not so shy as they are during
the day. I know an expert gunner who has sculled his
punt through a large flock of coots at daybreak when he
was on the look-out for fowl, and they never arose, though
they kept turning and looking at him, sitting on the water
with their heads "reined" out. Some were feeding, fairly
springing out of the water, diving head-first downwards
and heels upwards, in a more clumsy manner than a wild-
fowl.

When the gunner does shoot him, he requires much pre-
paration for the table, for altogether "he is a bewty," as
they say in Norfolk. The Broadsmen, who eat him, pluck

him and dust powdered rosin over his down, and then pour boiling water over him, when his flesh is said " to come out as white as a barn-fowl's."   I have never tried this method; but I have curried him, when he's not so bad.

II

A Coot-Shooting on Hickling Broad

Some years ago, before a "learned" judge ruled—and so made himself a laughing-stock to the Hickling peasantry, gunners, marshmen, and wherrymen—that the tide does not ebb and flow on and off Hickling Broad, the natives, though not "experts," knew it did ebb and flow; and they know it now — knowing, too, that water may run up a river nine hours, or run down a river for $9 + 9$ hours, and yet the water will rise and fall; but then, to a "learned" judge such things are strange.   I repeat, before a judge and experts found there was no tide on and off Hickling Broad, the natives used to hold an annual coot-shooting, for Hickling is a favourite haunt of the coot, both human and feathered.   Let us take a hand in the sport.

It is "Christmas-Eve morning," the broad is still open, and plenty of coots are sitting on the open water, peacefully as india-rubber fowl.

For many days past every one possessing a gun of ancient or modern build has been furbishing it up, oiling the locks, cutting wads, and making other preparations; for a notice has been posted for some days at the water-side tavern that there will be a coot-shooting on "our broad" at 11.30 A.M.

Long before the old church clock has struck eleven, crowds begin to collect at the "Pleasure Boat"—all crushing into the tap-room, and calling for jugs of mild—whilst carts keep arriving; finally, the little green staithe is gay with laughing men and youths. Soon all the party has collected, some ninety persons, of strange dress and stranger accoutrements. There is much talk, and joking, and cheering as they crowd into the open boats—pleasure-boats, old cobles, marsh boats—some propelled by oars, others by quants (poles). So the chaffing flotilla of forty boats, for many have brought their boats over-night from mill-outlets and distant broads and meres, goes shoving and rowing off on to the broad, whose hundreds of acres of water gleam and ripple in the cold morning sun.

"Going ter shoot anything to-day?" shouts one man.

"Ay, bor. Shoot you—shoot each other," cries another gunner, ramming home a load of shot.

And so the fleet moves off, some of the boats carrying one gun, some two, and some even three; but as they push out on to the broad (the women and children watching them from the cottages on Stubb shore), their voices die away, for there are the coot sitting quietly at the other end of the broad. Gradually the flotilla forms into a hollow square, stretching across the broad, the right arm of the square being longer than the left. In this formation, which soon becomes a semicircle, the quanters and rowers move steadily towards the fowl. As they approach, the birds begin to move restlessly; and as the right and left arms close in, the rear of the formation keeping well back, a lot of the birds rise and fly down the deadly blind alley, when the air is rent by volley after volley as they move up the alley and cross the rear-guard, whose line has no end. Bang, bang, bang go the hindermost guns; birds are falling on all sides with loud splashes into the water; other flocks come and follow their friends down the deadly gauntlet—forty birds falling at a volley sometimes. There is much shouting and laughing as a dead bird flops on to a quanter's head, or a man's face is blooded; and, finally, when the last bunch has flown over the puntsmen's heads, all hands seize on the birds, throwing them into the boats. Whilst their living relatives are re-settling behind them at the other end of the broad one man on the right wing is calling out that he got ten shots at them as they passed, whilst another is complaining that he got never a shot.

When all the birds have re-settled, the gunners re-form in the same way, and start back again, growing silent as they approach the birds, that are more restless now, and sooner take to the wing, flying over in companies in the same manner as before, and meeting with the same end. Perhaps in the middle of the hottest fire some man will yell out that he has been shot through the leg, or got a shot in his head, for in the excitement some shots go amiss, though I have never heard of any one being killed at the sport.

So the party works backwards and forwards five or six times, when, perhaps, there isn't a coot left on the water, all that have escaped having fled to the adjoining marshes or covers, and mayhap, too, the boats have lost their formation, and are scattered all over the broad, some with piles of dead coots in them, others with scarce a bird, and then the tally begins, crew shouting to crew—

"How many ha' you got, bor?"

"Oh, I ain't got one," says one man despondingly, and perhaps a boat near by teases him in turn with a "There you are, old Poker."

Another crew will shout gleefully, "We got twenty-four. Any one beat that?"

No one replies.

So the same voice calls again cheerily, "We're head boat, then."

Then you see the scattered flotilla making for the inn, some buying coots on the way, others begging for them, others promising drinks for "a brace."

And soon the short winter day is over, and the noisy

crowd at the inn disperses, making their way through the dark lanes and muddy roads for home; and the poor coots collect on the hovers, and try to sleep over their experiences, or else rise in a body and seek another broad—Horsey Mere, mayhap.

COOT'S NEST AND EGGS. (*In situ.*)

CHAPTER LXXXIII

RINGED PLOVER

Is not uncommon on the mud flats of Breydon and round the coast.

In April, as you walk by the sea, bordered by the shifting sand-dunes, fringed with marram, fortalices that protect the flatland from the sea, you will come across these birds feeding in the pools left by the tide on the shifting shore; and if you leave the beach and wander over to wind-sculptured galleries decorated with dry marram roots, you may in some cosy hollow, where the gravel lies thick upon thorn-bushes, placed there to strengthen the sea-blown sand, come upon the ringed plover's eggs, placed on the finer stones.

There are generally three of these eggs, and they are difficult to see, even when pointed out to you by more experienced eyes; but nowhere are they common in this district.

You will always, too, find a bit of sea-weed near the eggs.

Later in the year, too, when the marsh-mowers' voices sound over the sandhills, you may find the stone-runner's eggs, for they rear two broods in a season.

As the sun gains power, and the bright hot days of July beat down on the gleaming sandhills, you will, as you wander by the marram-fringed sea, come across little flocks of these pretty birds flying from pool to pool feeding, calling

as they rise and fly down the beach before you, alighting fanwise on the yellow sands. That is the time to shoot them, for they make a capital dish, and taste nearly as sweet as a snipe.

RINGED PLOVER'S NEST AND EGGS. (*In situ.*)

CHAPTER LXXXIV

GOLDEN PLOVER

OCTOBER gales bring the golden plover to the Broadland, gales in which this bird, like the peewit, delights; for you may see them playing in a gale of wind, flying up and dropping, enjoying themselves as a good sailor enjoys a stiff breeze—when his ship is all alive. The golden plover is a marsh-bird, for there he finds worms, as does the peewit —food on which he lives and grows rich for the table; for there is no better bird on toast than the golden plover.

On the winter marshland you may see large flocks, sometimes numbering two hundred, feeding quietly with peewits, but never scattered about, generally eating in close clumps.

At night they flight, flying low and swiftly to the land. By day, too, they will flight, whistling as they go, a curious whistle with a break in it, like a toy-whistle; and you may on such occasions see them with necks well drawn back into their bodies, whizzing through the clear blue air, often flying over your boat.

When they are feeding on the green marshland, they will at times allow you to approach quite close, for they are filled with curiosity, and are therefore not difficult of approach. The gunner knows this, and walks down to leeward, as if he were going to pass them unmolested; but he does not, "monster" that he is. The gunner, too, makes more capital of this curiosity; for he knows that, like many fowl, the golden plover likes a dog, so he turns his dog out on to the marshes, and they draw up to the dog, following the beast as he returns to his master, who lies hidden behind a shoove of reed. They will often be decoyed within shot by this simple stratagem.

I should think many of them could be caught in a flight-net at dusk, for that is their favourite flighting hour. For though they whizz past at lightning speed, they generally fly three or four yards above the stuff. Indeed, in hilly, uneven countries, hundreds kill themselves against telegraph wires.

When they flight by day, generally from north to south, or the reverse, they keep going in small bunches all day, and a knowing gunner gets in their line of flight and kills many.

In autumn or early spring they frequent newly-planted wheat-fields, or fields whence wheat has been harvested; nor are they easy to be detected in such a place. With the first fall of snow in the marshlands they take to the uplands; and if the snow continue, they disappear from the district, generally going south.

In April they change their plumage, and do not look like the same birds as you see them walking about the ploughed lands with their black mails. But in May they have gone from the Broadland—gone to nest elsewhere and bring up their young, to return again with the October gales, and reappear on toast amid the chrysanthemums.

CHAPTER LXXXV

THE PEEWIT OR "PIWIPE"

THE mournful-voiced bird of the marshland is the crested lapwing, with beautiful plumage, but without much attraction beyond—a bird giving one the impression of complete selfishness and pride. Such is the piwipe, a bird dearer in the observance than in the character, but most desirable upon the soft-lighted dinner-table, provided he be not shot on the beach, where grow the bitter grasses. But let us to the marshland.

It is a bright day in February, the landscape is brightened by a glowing sun seen through a warm south-easterly wind that blows the delicate waves of shimmering heat across the newly-cropped marshes stretching away to the rustling reed-beds and gleaming sand-hills, beyond which is heard the hoarse roar of the sea.

As you sink into the pale dead grasses upon the fringe of the marshes, the azure air is suddenly dappled with vast flocks of birds, that you recognise by their flight and voices to be peewits and golden plover: starlings too are flying through the liquid blue on their own account, and farther on the right a flock of linnets scurry by chattering.

The piwipes have not paired yet, for the season is dry, and the peewit delayeth nest-making till the marshes are soft. They like moisture and water near at hand. They have no fixed day for pairing. You may see them in large flocks describe many-shaped patterns on the blue one day, and lo! the next the flocks are scattered and the birds paired. Courtship is a short, very short, period with them, which makes me think the peewit, and many another bird, pair for life.

But the spring is backward, and the wind blows dry from the icy east. So the flocks keep together, feeding upon the marshlands by day, and flighting to the uplands by night, flying low above the reeds and dikes, where the flight-shooters lie concealed; and as the tongues of fire from the guns shoot upwards across the purple, the birds turn and lap and bustle—the leader crying his melancholy note, and lo! they are gone, a few of the flock being left dead on the cold water, wherein the stars shine like bright flowers. And they are one of the most difficult birds the shooter knows when they begin lapping with their rounded pinions.

And by night they stalk the upland feeding—especially upon moonlit nights—over the newly-turned earth, upon worms, slugs, snails, and insects, doing the farmer good work. And when the daylight shines in the purple skies, they generally go back to the marshes and dikes to wash their mud-dabbled feathers; and then you may see them standing on the wet grassy shores strutting and preening their feathers, crying out to each other as they wash the earth from their feathers, and then to breakfast—perhaps at the tail of a marsh farmer's-plough—with rooks, starlings, and, if late enough in the season, brown-headed gulls, snatching the worms from the glistening soil; or mayhap they go to the sea-beach, now hot and teeming with life, for the tide has withdrawn, and their table is spread; or mayhap to the mudflats of Breydon, where the worms wriggle in the rosy-tinged ooze. But they prefer the marshland, and love stagnant dike-water for their looking-glasses.

And when the lengthening days of March have warmed the sandy warrens, some of the cocks in the flocks frequenting that district begin to tumble about amongst the hens, calling, "Three bullocks a week, week arter week;" and the fenman's heart is glad, "for they'll soon be laying now," he says with bright eyes, thinking of the six shillings he will

S

get for the first dozen of blotched eggs. But his heart is gayer still when he sees both birds sitting about on the warrens, and mayhap on the ploughed marshes as well as the clear marshes, for he knows the beginning is near. He saw them tumbling nearly a fortnight ago, and he knows they generally lay three weeks after they begin tumbling, or "pairing," as he calls it. But when he sees the cocks fly up and cut at an old grey crow that has just flown over, he is assured "there be eggs," and he is right; we should find eggs.

But let us select our marshes, for we will not go to the warrens, although the first eggs are sure to be found there; the soil is warmer there. We will go and look over a clear marsh, a ploughed marsh, a grass marsh, or a new-lay, and a few days after the first eggs have been found, for we wish to see the birds busy at their great task.

It is a beautiful dawn in early May, the daylight sky brightening to the nor'ard, as we start in the heavy dew up the wall; for daybreak is the time to find a duck's nest, and soon after daybreak a peewit's cradle. We will go down now across that dike into the marsh, where the cocks of litter stand piled, ready to be poled to the big marsh-boat, and carried to the farm. As we walk across the dike over the old plank, all riddled with bolt-holes—for 'tis a footbridge torn from some wreck salvaged from the Hasboro' sands—I throw my cap into the air, and look! See yon bird silently and swiftly flying off across the water and away over the reed-beds? That is the hen. You must watch *her*, and her only, if you want to find eggs. But here comes the cock tumbling, and excitedly calling, "Three bullocks a week—week arter week—week arter week." An expert egger could tell you how many eggs she has by her flight, for as her eggs increase her flight gets more sluggish, and when she begins sitting, she is "a real old lump" when she flies away, and indeed such is the case with most birds. But the hen-peewit is exceptionally active until the second egg is laid, which, by-the-bye, is not the day after the first is laid. They sometimes lay every day— sometimes omit two days without laying. But we will pay no heed to the old cock; we know his tricks to lead us away on a cock-peewit chase, which is far worse than a wild-goose chase—for we are very likely, if unwary, to attempt the one, and the other we should not.

See those shallow cup-shaped depressions near those old thistles. They are cocks' nests. He begins the game by "scrabbing" several of them. I have counted five such made by one bird. Some here, you see, are lined with pulled grass; but they are nothing. *She* builds the real nest, and that is why the fenmen are more delighted when they see her on the ground; for they know that when he is on the ground alone he is only pretending with his "cock's nests." But you see the cock-bird has gone too, since he could not take us in. But no doubt we shall flush more birds on this marsh, for they are sensible birds, and often build in company with each other, and snipe and red-legs too build near them on suitable marshes; but this one is too dry for them. But some couples are unsociable and drive off all socially inclined couples.

Look, there is an old cock standing by yonder fork left standing in a heap of litter; the hen has flown silently away whilst we have been talking. There are eggs thereabouts, for he is watching them, as the cock always does when the hen leaves the nest. But see, he is up and coming towards us, crying the usual cry—hovering close over our heads, as if eager to pierce our caps. There must be young; he is so excited. Had we a dog with us, he would dart down almost within a yard of it. Let us search here amongst these thistles and rushes, and be careful you do not tread on the nest, for it is very easy to pass it over a dozen times. But see, here it is, just by this thistle-stalk, with one help- less but pretty little chick, two of the eggs "sprung" and the fourth egg sound, but bright, smooth, and warm to the touch, as all hard-set eggs are. The eggs are lying in the grass-lined cup on a slant, their small ends pointing inwards, and that is their usual position, though the hen turns them round every day, as, indeed, I think most birds do, so that they may be evenly warmed into life. The pretty little fellow is evidently only just born, or he would be out of the nest, crouching in the print of a cow's or horse's hoof, and the eyes of Argus alone would find him. Once only did I find a youngster out of his nest, and that was on a bare mountain in Perthshire: there was no cover. This youngster here, had he been old enough, would have run out of the nest, and, once having left his cradle, nothing would have induced him to return. But he is only just hatched: for had the four chicks been born, both old birds would have stayed to defend their young; by that sign shall you know whether there be young.

But let us lie behind this heap of stuff over here and watch the first hen we flushed; she ought to be returning soon, for I have seldom known them leave their eggs more than twenty minutes, if there be more than one egg. But she is not sitting yet, or the cock-bird would have stayed longer; and she begins to sit when the third egg is laid, though four is her full number.

Let us wait here, though a drizzle has come on, greying the trees and distant sandhills, for the birds are sure to be back soon now; for a peewit *never* leaves her eggs exposed to the wet. She always covers them, or more usually returns at all hazards and sits upon them. But hist! there she comes across the wall and flies down to the marsh. See her. She is looking intently. "All's well," she thinks. See her running along for a few yards—for peewits seldom walk, but run, after appearing to pick up something as they go. Then she stops again, and listens, and on she goes right on to the nest by that heap near the dike. Soon you can scarcely see her with the naked eye. And yet these brave birds sit upon the lonely marshes through the night-watches, regardless of the cruel and fierce rats and ruthless weasels and stoats that are breeding near by, and who often rob their eggs and young, if they do not eat the mother herself. And yet the birds sit on through darkness unprotected, merely obeying an instinct stronger than fear. Having made our mark, after the manner of the fenmen, we run up and flush her. She rises again, and flies lapping off, silent as death, striking away over the water again; and after a little search we find the nest, lined with rush and broad-leaved grass, upon which lie two eggs, heavy and sweet and fresh, as the water-test in the nearest dike proves. And we will leave them, for they are not so good to eat as a fresh hen's egg, and chance the old Kentish crows sucking them, as they often do, as well as eating the young peewits when they can catch them. But see, here comes the excited cock again, tumbling about, and thrashing the air with his wings, and, till he is about to fly up, there he goes into the open, twisting, and turning, and shrieking his dull refrain, " Three bullocks a week—week arter week—week arter week."

And so these birds lay on, if robbed, unto their clutches of four eggs each, or a round dozen per bird, laying as late as harvest-time; for they must have a young brood if possible. But they leave the place their early nests are robbed in, and go moving from marsh to marsh at each new loss of embryo; and so regular is this retirement before the eggers, that the experienced say, "Ah! well, they'll be inter our marsh next, directly they're robbed over yonder." But this egging has sadly thinned their numbers; and instead of being able to find two dozen nests of a morning, as was formerly common, we may be lucky to find one in many a marsh. And when the young are hatched, the family seeks a slad or soft marsh where the water lies, or the ronds by the mere-side, where they lead their young, decoying them by cries, as do the red-shanks, but seldom making much of a journey,

for they nearly always build near the water. And when
they arrive at the feeding-grounds, rich with snails, grubs,
grasshoppers, slugs, worms, and small shell-fish, the young
spend their time in feeding, and the old birds fly overhead,
when not feeding too, calling loudly upon their young, and
directing them when an enemy is on their trail; for the
harrier and crow, the rat and stoat, often prowl through
the rushy brakes that the young frequent in search of
a supper; and sometimes on a moonlit night you may
see these white birds flying and turning over a rond, and
filling the peaceful night for hours with their anxious cries,
for the voracious stoat and rat love to work by moonlight. And your heart is full of pity for the heart-broken
parents and the innocent youngsters crouching low in some
pool or hollow, fearing every moment to see the sharp, erect
ears and glittering eyes of their tiger-like foe; and often a
short child-like shriek tells you one of the flock has gone
for ever, ere its life had well begun. And as they batten
and fatten on the succulent life of the swamps, they soon
grow as large as their parents; and then, before they can
fly, and when they can fly they do not make use of their
newly-acquired powers before danger, but will squat low,
as they did in childhood, so powerful is the force of habit.
And, indeed, at such seasons they seem peculiar, and let
you approach them pretty closely, merely running along
and pecking at the ground, as though feeding; for they fend
for themselves before they can fly, and at times the old
couple leave them when they can find their own food, and
before they can use their wings.

Directly they get full use of their powerful wings, they go
to the marshlands, preferring a grazing marsh, for that is
generally dry; and they always choose a dry marsh to
sleep upon, as it is also rich in food chosen by the birds,
and animal remains. There they spend their young life
by day and night with some old birds, though at times
a few of the young wander farther afield into a turnip-

field—dear to them. I have seen the young begin their
flightings or wanderings as early as the eighth day of
July, flighting to the uplands at dusk, and returning to
the water-side at dawn to wash themselves and feed, thus
reversing the usual order of water-flightings. And at this
season, if you are fond of the fens, you shall know the old
from the young on the darkest nights by their voices—the
old birds' call being mellower and more rounded; the young
scions of the flock giving hoarser voice. And you shall hear,
too, the leader—for they always have a leader when flying
in flocks—who is an *old* cock, full of experience and wary.
And when the equinox fills the North Sea with cod, fresh
flocks come over the rain-swept seas; and even in the marshlands the air is full of flocks of peewits, starlings, rooks,
and golden plover, swept across the sky in ever changing
patterns by the fierce gales. And these, after Michaelmas,
too resume their evening flightings to the uplands—once
more being turned up by the plough—and feed upon the
worms, returning at daybreak to their beloved marshes
or the cold sea-beaches; but they very rarely go to the
beaches if there be good marshland between the upland
and the sea. And at this season the fenman is ever on
their flighting line, and many a "piwipe dumpling" is
cooked by his goodwife, after she has laid the "bahd" in salt
and water for a night—and thus is the Englishman wiser
than the Scotsman, whom I have often seen turn his nose
up and draw down the corners of his mouth in disgust when
I talked of shooting and *eating* a *tuchet*. Yet we used
to "size" for them in the old days at Cambridge—"green
plover—9d."

But soon the harsh breath of winter 'gins to blow over the
flat land, and the birds grow restless, moving about the marshlands, an augury of coming frost; and the fenman's heart is
happy when his friends do not leave, for the augury assures
him the frost "ain't going ter last." Nor does it; for these
sensitive living barometers seem to know full well when hard

weather is coming from the frozen north; for as the white outriders of an Arctic storm sweep across the North Seas, these birds arrive in great flocks—describing a beautiful decorative pattern upon the blue—and flying towards the stars, they start off at the leader's signal, and go where frosts are not, seeking a serener clime. And the marshland knows them no more until the weather breaks up, and the frozen rivers, dikes, and broads are loosed, when they suddenly return to their pastures—always new. Thus, if the winter be hard as that of 1890–91, for months together never a piwipe is seen until the ice breaks up and the amorous frogs begin to croak in the swamps; then they all come bustling back to the sere marshes. But it is late ere they begin to tumble and make cocks' nests on the flats. Sweet is their melancholy cry resounding in the mutter of the sea—a voice that should be protected and encouraged, for the farmer's sake, to reproduce itself; but the citizen, with round belly, has eke taken a fancy to a crystalline white and a mealy yolk enclosed in a dark brown blotched shell, and so the birds are robbed, and so the native piwipe is being wiped out of the land, and the immigrants do not taste as sweet as a young native. Ask any gunner, or try yourself; for a good plump young native bird of the year is a treat too good for any *gourmand* of plovers' eggs; 'tis a treat for a *friand*, and may you be one.

CHAPTER LXXXVI

THE AVOCET

THE day has broken cold and raw. Still you go down to
your boat, and push off into the cold grey tide running
fiercely to the sea, and keep yourself warm as you row
along the saltings, flushing "ox-birds" and "stints," and
here and there a flock of widgeon floating on the shallows,
for the tide has left the flats, and the slub is dripping wet and
drying in the cold morning, as you paddle quietly past a grey
salting startling a flock of curlew, that go whisking away in
the grey. And suddenly the curious *t-klöee* of the avocet
catches your ear, and you look up a drain and see the dignified bird standing some hundred yards away ; and you know
'tis an avocet, or "shoe-horn," as the old Broadsmen call him.
He knows he is safe at that distance, mayhap, for he does
not rise, but goes on gently picking the worms from the
slub, now pink in the rising sun ; and you too go on your
way, having caught sight of a bird now very rare, because
the region of the marshland is contracting, and the collector
—a plague on him—is increasing.

CHAPTER LXXXVII

WOODCOCK

IN the October moon, when the reed-cutter has begun his cold work, standing amongst the icy crops with his meak cutting and binding the yellow sheaves of reed amid the roar of the sea, thinking of his old herring-fishing days on the North Sea mayhap, the swift-flying woodcocks come over with a fair east or north-east wind and a moonlit sky—for these birds always prefer to see their way, which may perhaps account for so few of them colliding with lightships or the garish lighthouse windows. And when the marshmen go to their work in the chill mornings, they come upon the woodcock close by the sea, in the sandy roads, on the marshes, in the stubbly uplands.

The gunners' flashing guns resound on all sides above the cry of the sea, for they arrive in fair condition, some indeed being quite plump; but the woodcock, as soon as rested, take to the shelter of the many-coloured plantings, where the cry of the jay, or the croon of the pigeon, and the challenge of the cock-pheasant is still to be heard. These grotesque-looking birds often make their way from the sea-beaches to the upland coppice under the cover of the hedge-rows—the marshmen as they go to and fro to their work knocking them out. Indeed few woodcock are seen on the wing down by the sea. They are a skulking bird, and should you see one sitting under the lea of an autumn hedgerow bright with scarlet berries, if he see you, he will squat, and nine times out of ten, the gunner—who knows this trick of his—can walk within shot, and the travelled woodcock soon lies dead at his feet. He knows, too, that he is a-hungry on his arrival, and loves the turnip-field, where the luscious grubs and worms grow fat and juicy, and thither the gunner goes "afore he takes ter the woods." But when he does fly, as he does at the flighting hour, the gunner knows him as he comes along in the growing winter dusk over the reeds; for he flies about five yards above the reed, taking a bee-line, and flying swiftly—something like a peewit, though they turn about a little as they approach a slad, where they alight to sup, but not so with the woodcock, he travels straight ahead, unless he be knocked over by the alert gunner, who stands in his grey punt in the shaking reeds.

And through the dark and dreary winter he is scarce along the coast. In hard winters he usually disappears; but upon all occasions he is rare in winter, one gunner considering five a good number to bag in winter. But his scarcity may be due to his skulking habits by day, for he is really a night-feeder, working his wings and boring like a snipe into soft marshes, whither he flights at dusk from the plantings where he dreams the day away.

And when the trees are bright with bursting buds, he begins his return journey, and his hoarse snipe-like voice is heard as he falls a victim to the gunner, many of whom declare he nests about the district in osier carrs; but as they never find his eggs, this may be doubted. They think so, because they often find the females with well-developed eggs—"full of eggs," as they express it. But so are many of the pochards that are shot. Yet a pochard's nest would be a "great find."

Truth to tell, the eggs seem to be well advanced in many birds at the time of migration, and perhaps this is the great stirring in their nature which impels them to seek a more congenial clime, where there will be plenty of food for their young.

And so, like many another bird, we get glimpses only of

the woodcock in life and in death.   Good as he is on toast, he is, in my opinion, surpassed in quality and flavour by many and many a bird of lesser reputation—the common waterhen and water-rail both being finer table birds, I think, to say nothing of snipe, golden plover, and many another bird.   But the world is full of over-estimated birds as well as false reputations; but the truth is great—it will prevail.

CHAPTER LXXXVIII

SNIPE

THESE grotesque birds, all belly and bill, were born for the table—a long bill to feed a dumpy, graceless body, which, however, is as sweet as a nut, and more juicy.

Rarely at Christmas-time, but more commonly when the young reed and lily leaves and gladen spikes are just peeping above the bistre water in the dikes—when March, the wind carrier, is with us again—when one hears, high up in the blue sky, the drumming of the courting snipe, the male bird, who flies up to the pale silver crescent of the moon, and, turning on his side, drops in the blue liquid air, drumming as he descends for a short flight ere he gathers himself together and ascends with a coarse chucking voice ; **flying** round and round over the moonlit grassy marshland, his flight extending a quarter of a mile, courting his sober, serious mate with the long bill and solemn mien. And the origin of that noise is still wrapped in mystery; but methinks it is a vocal sound, a voice propagated through the closed sieve-like membrane wrapped round his long mandibles. Nor do I think his wings and tail take part in the strange music, as is always suggested. Indeed, 'tis easy to imitate the snipe's drumming by closing the mouth and covering the nostrils with a handkerchief, whilst uttering a drumming sound away down the larynx.

"Summer lambs," the fenmen call them, on account of these strange bleatings, noises always uttered by the cock-bird. At this season, too, they delight in playing over the sere reeds ; and the fenman's heart is gladdened, for 'tis, he says, an augury of fine weather, even though it is showery at the time, and they drop into the reed after their flight as if all their bones were broken by the ordeal.

At this early season of the year (say in April), you may hear the snipe and red-legs drumming and calling all day, and by night too, and often in company, for they both love the moist marshes, where the beds of cotton-grass and soft grass grow. Indeed, they often nest in company, they two and the green plover. "One coys the other," as the fenmen say. And I have found their eggs in the first week in April close together.

And their simple nests, just a cup-shaped depression in a tuft of dead grass, the grass being drawn together in a bunch over the top, is all the trouble either snipe or red-leg takes over its cradle. Once, 'tis true, I found a nest carefully lined with dead rush, but such is rarely the case. There they lay their four large blotched eggs, astonishing in size, too, compared with the bird's size, and most delectable for eating, far surpassing a peewit's. But the birds are far more delicious than the eggs, which should be left untouched for a fortnight, when the pretty little snipes will be all alive in the simple nest. Should you go into a marsh in early spring and disturb the hen-bird, she will run off and rise, flying and scaping; and if the eggs be glossy and light, she'll fly round you scaping piteously. Throughout the period of sitting, the male bird drums overhead, encouraging the patient hen, and especially at dusk, when a soft shower of rain has powdered the bent grass, the cock is most active with his bleatings. Indeed, after a rain-storm they are all alive, drumming and resting by turns, sedately and grotesquely upon an old mill or bare tree-top.

When the big old hen starts off the nest—for in the laying season she is the larger—to feed on some bright spot of water, about half-an-inch deep, lying on some puddingy marsh or eke upon a hard-bottomed flat—they like both —provided she has not been previously robbed by some egger, if you follow cautiously, you may see her drop to her feeding-grounds from the sky with a *chuck, chuck,* and go bibbling with a hoarser *a-chuck, a-chuck, a-chuck,* thrusting her long bill into the duck-weed, of which they are very fond, stirring it all into a froth, so quick and eager are they to snatch the worms and weed—and you know where they have fed by their liquid dung. Then, perchance, on a sunny day, you may see the cock lie on the grass sunning himself, with outstretched wings, like a turtle-dove. Indeed, you may watch all this at leisure, for snipe can be kept in confinement. One gentleman I know kept a number of snipe and jack-snipe, and he tells me he always fed them upon a mess of worms, duck-weed, and water, the birds eating quantities of worms each day. And the same gentleman tells me he once flushed a jack-snipe and a whole snipe together; so they are good friends. Though preferring the inside marshes when wet, you may find snipe in the most unlikely places. On stubbles and on turnip-fields flocks of thirty or forty may be seen at times, provided there has been enough rain to soften the earth. Again, if the weather turn hard suddenly, they go to drains and dribbling brooklets, or locks still unfettered by the ice; but when these last water-oases become frozen, they go away to more congenial climes.

After the young are hatched, if the marsh has become dry, the old bird will lead the young to the water-side, and there you may see them feeding by the dikes; and towards the end of August they are to be found about the inside marshes and on the ronds in clutches of eight, ten, and twelve in number. At this season, too, they begin flighting at dusk to their feeding-grounds.

And a pretty exodus is this to the water-side. Of a still summer evening you may hide in a sappy reed-bed just alight with the last rays of the setting sun, and lo! suddenly you hear a snipe *a-chucking, a-chucking* through the silent eventide; and if you be lucky, you may see the little procession, the cock and hen flying ahead of the beautiful

quaint, timid little nestlings, calling *a-chuck, a-chuck,* and
alighting a yard or so ahead on the marsh, the young ones
running eagerly up to their parents chirping; and so the
pretty little family goes its way to the spongy water-side,
where, under the rising light of the young moon, they take
their first lessons in bibbling. There they stay by the water-
side till they can fly, hiding like a mouse in the rank marsh
crops if a hen-harrier darken the sky, for that old bandit
loves a young snipe as well as any *gourmet.* And then the
family ranges the marshes, and a few friendly families join
company; and by shooting-time herds of thirty or forty fly
restless in wet weather from marsh to rond, or from rond to
marsh. They won't "lay" in wet weather, for fear, mayhap,
of wetting their feathers, and such is a bad season for the
gunner.

Should they have been robbed of their huge eggs, they
will lay again, so that young fledglings are at times found
as late as harvest—a fact the gunner regrets, as he walks on
a still, bright day down to leeward, never to windward, ever
ready for their turning to right or left of him, affording a
splendid mark; for if you go down to windward the snipe
will rise and twist away from you, offering a poor mark.

And at that season they are plump, as an acute lawyer
once found. An old gunner had been out snipe-shooting
with him one day in August, and on the way home through
the straggly village the gunner made excuse to drop into a
shop to buy some high-dried herring for supper, when the
heaviest snipe was weighed—six and a half ounces.

On rejoining the lawyer, he offered to bet on the weight of
the heaviest snipe, backing it weighed over six ounces. The
lawyer said no—under. The money was staked and the
snipe taken to the post-office, and lo! it weighed over six
ounces and a half; and the old gunner is happy to this day
that he bested a lawyer.

This same old gunner remembers when it was quite easy
to kill eleven brace of a morning—twenty years ago; and he

often tells of a shot that had travelled one hundred and fifty yards, and many such details dear to the sportsman.

When the alder-carrs are embrowned in October, the migrant snipe come over in company with the woodcock, both carried on the same propitious breezes. And at times, if you be wandering on the wind-sculptured dunes, you may see these birds arrive and drop dazed on the wet beaches, giddy with fatigue after their long flighting.

A strange bird is the snipe at such seasons. One day the marshlands will be alive with birds, and the next they will have gone; the floods were out, or the wind blew too keenly from the east. And the way they vary in plumpness is astonishing; one week they will be as fat as a capon, another a mere feather-bag of bones.

THE JACK-SNIPE, OR HALF-SNIPE,

Is a splendid little fellow on toast, especially when he weighs four ounces, as I have record. He, too, comes to us in frosted October, spending the winter and spring, staying on until the hawthorn hedges are green with the new leaves, and rarely all the year round. The half-snipe is best for the toast just before he goes in the spring; then a good fat hen, which is the larger bird, is *sabroso*. But he is shy, and is not fond of betraying his whereabouts, though he scapes when flushed; indeed, he will allow you to step within an inch of his little body ere he will budge—indeed, a knowing fenman once caught one with his hat, so that without a dog many a gunner walks over him time after time. So the little fellow is wise to draw himself down to the ground, lay his back out straight, his head down, and sit as if he were asleep. Yet he keeps a sharp eye on you; and if he deign to fly, his course is short—he soon drops into rushy cover.

He dearly loves a floating hover by the mere-side—a land that rises and falls with the tidal ebb and flow—a place that

is always moist.   And upon a small patch of such no man's
land a dozen often congregate, and feed and sleep the live-
long day.   At times he is more adventurous, as I have said,
accompanying the whole snipe; but he is at all times hard to
shoot if not allowed to steady his dainty little morsel of a
body; and in a breeze of wind he is well-nigh impossible.

On a cold night in autumn you may walk down to a spongy
marsh, and wait, perhaps, for the flighting of wild duck from
the sea, half-buried as you are in a rush knot bordering a
slad, suddenly, in the growing chilly dusk, you may be roused
by a rustling of wings, and a flock of jack-snipe dart suddenly down into the slad.   They come and go like restless
spirits.   The night seems to be their favourite season for
ranging from marsh to marsh.   Often in the dark winter
evenings, as you hide in an old eel-boat, lost amongst the
yellow reeds, you may see a jack flighting across just above
the reed tops, or else flying high overhead near the winter
stars, but seldom in pairs.

Though shy, the wily fenman traps the jack as well as the
full snipe.   Stealing forth in the cold evenings, he searches
the likely marshes for spoor and the frothings of bibblings.
There he sets his springe, as "fickle" or "slim" as he can,
just underneath the water, having first "riled" the water, or
stirred up the dregs to foul the pure surface, so that when the
water settles and purifies the fine silt will settle on the trap
and hide the iniquitous "steel-fall," or common steel rat-
trap.   And many a fenman gets four or six birds in this
unholy manner between the winter evening and the winter
sunrise.   Still they are good on toast—*que voulez vous ?*

The Great, Solitary, or "Solingtary" Snipe,

As the fenmen call him, is a rare bird now-a-days in the fens,
though formerly reported to be common.

He is a cunning fellow, and never betrays his succulent
carcase by scaping sillily when he is flushed from the soft

places **where** he dreams in seclusion; but still his plumpness, greater spread of tail and characteristic flight—like
that of a golden plover—and the two external white-tipped
tail-feathers, betray him. He, too, is excellent on toast,
though hard to come **by**; for he is "more slinky," as the
fenmen say, and rarest of all.

And so farewell, my grotesque birds, dedicated to **the**
chef-de-cuisine—fare you well, with **your** monstrous love
songs, strange guzzling habits, amusing skulking flights.
You, too, have your poetry, after twenty minutes in a hot
oven—on toast; and may you yearly bring forth your
beautiful little fledglings, and lead them safely down to the
bright waterside for aye—this baptismal pilgrimage being
but the forerunner of the great sacrificial feast that awaits
them on the table of all who are artists enough to know a
snipe from a hernshaw. Increase and multiply, ye grotesque
ones—and **may the** *chef* be with **you.**

CHAPTER LXXXIX

THE COMMON SANDPIPER

IN early spring, the voyager through the waterways of
Norfolk will often hear the peculiar *t-learow*, *t-learow*,
t-learow of the common sandpiper, or the more simple-like
a-chuck, *chuck*, *chuck*; and as his sail brushes past the river
grasses, he may flush from some dike or rond the white-
breasted sandpipers, who love to feed by the waterside.
But they are quick birds, and flash against the blue for a
moment and are gone.

And again, in late summer, he may startle them from dike-
water, rich with sapphire colours; but of their habits he will
hear little, and of their nesting learn nothing. A mere
vision across the spring and summer blue is all that is
vouchsafed him, or the peculiarly melancholy and persistent
call through the night-watches.

BLACK-TAILED GODWITS. (*From life.*)

CHAPTER XC

THE RUFF AND REEVE

SOME fine showery day in the middle of April, as you sail through these sluggish water-ways, you may perchance see a small bunch of birds flying against the flecked azure—birds that you take at first to be red-legs; but, upon second thoughts, you notice their flight is swifter, and that they glide along somewhat after the manner of a swift ; neither do they call like the noisy red-leg. Then you know them to be a bunch of ruffs and reeves. Perhaps there will be four ruffs, with their collars half-grown, and twenty reeves. The old fenmen say five reeves to a ruff, but this may be or not. And perhaps—you with your glasses —you will observe that they are all reeves, and never a ruff.

You watch them fly silently down to a hill—a flat, bare, rushy place—for hills are *flat* on the Broadland. Putting your boat about, you glide back over the dark beds of hair-weed, startling a spawning pike from his warm lair, and running up head to wind, you lower sail behind an alder carr, shove your boat up to the shore and land, stealing softly along a wall that overlooks the bare stubbly marsh, their favourite "hill" that you read of; but you see no stations taken, no circles beaten, no eminence, and you know wel there's another fanciful "truth" exploded. But watch ; there are a couple of ruffs and eight reeves. See yonder on that bare patch. They have their immature grey ruffs blown out and they are dancing, whilst the sober-hued reeves look on They do not fight, but seem to dance round as if on springs, occasionally darting at one another, like young gamecocks or pigeons ; but there is never a blow given, never a sound uttered—'tis merely a rival exhibition given before the ladies. And every fenman who knows them will tell you what you have seen is the usual proceeding. Even if only one bird be present, he will dance as usual, not confining himself to any circle, but running down the wall or over the marsh, as his fancy listeth. And now you know another "truth" is no longer true.

How they arrange matters I know not, but one ruff takes unto himself many wives, serving them all. An old gunner once saw a reeve serve five wives "of a morning." He is a sensible polygamist.

In April, too, you may see, perchance, small bunches of these birds flying about the marshlands, disappearing for days and reappearing as suddenly, though they never all disappear, but the bulk go—it all depending upon the season—at the end of May, when the few remaining reeves make their nests, if the weather be fine, warm, and still—merely making a cup-shaped depression in a tussock of grass or sedge growing amid short grasses upon soft marshes near the water, where live the insects upon which they live. The nest resembles a red-leg's, with this distinction, that the reeve always has one or more circular runs to the nest, like a mallard, and pulls the stuff over herself, like a waterhen. There she lays her four eggs, precious gifts of the gods, resembling those of the red-leg, only darker and differently blotched, and "squatter." The difference is never forgotten once the eggs are compared, but it is difficult to describe.* And all this time the ruff has been growing his fine Elizabethan collar, with its dots of many-coloured "sealing-wax," each bird being differently marked ; so say the fenmen, who have shot hundreds.

And during the monotonous sitting period—twenty-eight days—when the reeve sits so closely that you may at times

* Reeves' eggs were taken in this district in 1890, and young seen in 1892.

step upon her before she will budge, while at others she
runs off her nest like a duck, disappearing in the stuff, and
flying up some considerable distance away, the ruff generally
keeps somewhere in the neighbourhood of his harem; and
you must watch *him* if you wish to discover the eggs—watch
his "trams" or "workings," in the early dawn, and look
carefully over his trail, and if you find the eggs, leave
them.

And when the chicks are hatched, he sheds his fine collar,
and I believe goes down to the beach or mudflats, leaving
his wife and family "to do for themselves;" the young birds
creeping about the stuff, following their mothers, and eating
insects from the moist grasses, as do the "stints,"—chiefly
little grasshoppers. And indeed the reeves do not seem too
intelligent, for an experimental old fenman, who wished them
to return year after year to the same hill, was in the habit of
lifting their eggs and replacing them by the eggs of the red-
shank, and he avers they always brought the young red-legs
up, and didn't know but what they were their own young.
He once turned the tables, and the red-leg brought the young
ruffs and reeves up. This same old fenman scouts the idea
of their fighting; "their collars be tew fine for that," he says
with a knowing twinkle.

And all those tasks are performed in silence—never a
sound is heard; and such is the consensus of testimony
amongst the fenmen: the mysterious bird seems to be like
the monks, to the end of showing their collars once a year
and producing more ruffs and reeves.

And when the ditches are gay with red-shield-seeded
water-docks, and the marshes are white with fringes of wild
celery, and the fields on the upland bright with scarlet
poppies, and the river walls decorated with carved thistles
and the purple blooms of loosestrife, and the October days
kill off the insects, the reeves and their families leave us
to the mercy of frigid winter—perhaps to dream of these
sphinx-like birds: whose silence, coquetry, dress, poly-

gamy is their chief distinction, and it is a riddle. But soon they shall have passed away altogether from these islands, and be as mysterious as the dodo to the future Broadsmen.

REEVE'S NEST AND EGGS. (*In situ.*)

CHAPTER XCI

THE RED-LEG

As the fenmen familiarly call him, is the embodiment of the spirit of the fenland—so dear to him.

In February, when the shrewd nor'-easters blow the dreary sleet and snow-storms across the sere reed-beds and frost-embrowned marshlands, the quick, shrill, melancholy whistles of the red-leg may be heard; for this wild and hardiest of birds delights in the snow-storm and rain-storm. No weather will stop that melancholy cry and rocking flight, or silence that *a-whew, a-whew, a-whew, a-whew,* as they alight, or the more musical and wilder *T-liie, T-liie, T-liie.*

And so through snow-squalls and sunshine their early flocks keep coming in from the North Sea, until April brings the yellow king-cups.

A little later, restless and eager, you may see them pairing directly they arrive. On a fine March morning, perhaps, you will see a fat female (for she is larger than the male), with her spotted breast, standing upon a bare marsh-wall preening her feathers, when suddenly from afar comes the melancholy whistle of a black and white cock-bird, who drops from the sky beside her, and begins piping softly, and bowing his slender body from his hips. She, however, is disdainful. Yet he perseveres with his *wheet, wheet, wheet,* and still she heeds him not, till, vexed, he darts at her, whereupon the coy maiden runs off adown the wall. And so, pricked with true love, the chase continues for short stretches along the wall. Suddenly she doubles upon

him, and a swift race ensues, for they are quick-footed.
Suddenly she rises and flies off across the marsh, when the
fond young lover's heart swells with anguish, his sobs become
double, and he bows his elegant body as if she was looking at him.　Then he, too, flies after her with loud pipings;
and perhaps in some shallow amongst the reed-stalks she
accepts his handsome red-legs, and they become one.

And some daybreak in early April you hear the pair calling from the big ronds or broad shores; and later, when
the sun bursts forth, you see them in the blue flashing over
the sere marshes; and if, perchance, a leaden sky creep up
from the sou'-west, and break into fine grey rain glistening behind the sunny foreground, you may see these birds
flying hither and thither, piping sadly as they revel in the
soft, dreamy bath, whilst the calves have left cropping the
catkined sallow tips, and hidden in a lair 'mid the island
clumps of marsh growths; and all the landscape is gleaming
with bright eyes, touches of light glistening from bedewed
eel-cabin, mill, or shepherd's refuge—eyes that seem to wink
at the joyous birds, and look shyly upon a wherry with
lowered peak, recalling a maiden caught in a storm, with
her dress drawn around her.

Soon the azure is filled with pairs of these wild birds, and
you may see them far away over the green sallow breakers
of this green sea, calling and hovering in search of a marsh,
one year mown, moist and full of tiny islets of grass, wherein
they can lay their speckled eggs.　Or on the broad, where
the frothings lie embracing the young gladen shoots, you
see them flying from one side to the other, almost kissing
the water in their flight, or alighting by the mere-side, bury-
ing their sharp bills in the soft mud in search of food.　And
should you startle them, perchance they jump up and fly off,
calling their flute-like *toodle, oodle, oodle, oodle,* changing
to a whistle-like *uhū, uhū, uhū,* as they recede into the
blue.

When showery April shall have run half her course, you

may find the nest—indistinguishable from a snipe's, merely a cup-shaped depression cunningly placed in a tussock of grass—often beside the nest of the peewit and snipe; for they nest in colonies, the peewit choosing the driest ground, the red-legs and snipes the dry islets in a soft marsh. When the four eggs are laid, they draw the grass together over themselves to hide their prizes.

There, amid the feathery cotton-grasses, ragged-robin, and fresh grassy islets, these birds sit out their incubating period. The young couples are very shy, and should an enemy approach, the cock-bird, hovering overhead in the blue, calls quickly, and the hen-bird rises swiftly straight off her nest, flying away piping hurriedly, and making a quaint noise with her powerful pinions, and there in the grassy cup are the four eggs, with their light shells and blood-red spots, lying with their big ends uppermost. The older pairs, however, are not so shy. Once when watching these wild marsh birds I flushed an old bird from her nest, and immediately dropped flat in the moist marsh stuff. She soon returned from the blue. A fenman, too, once drew up and captured a breeding bird with his hat. She sat in his cap, and would not fly away till he coughed, when she darted off across the marshes; but she, too, returned in a short ten minutes, and finally hatched her four eggs. Dogs often catch them on their nests. And, indeed, at times you may stand over them and watch them sitting as closely as a hen-pheasant, the grass braided over them like a shawl. But not always is the moist marsh chosen. The rond is a favourite nesting-place, as it is with the peewit. There, overlooking some sub-aqueous garden, brindled with shadows of the reed, amongst which the perch swim in and out, discerning perhaps the shadowy reeds, and gloat as they hang their glassy necklaces of spawn upon the water-plants, the red-legs sit for a sleepy fortnight warming the yellow yolks into the quaint little bodies that soon will fill the stuff with their baby cries, for seldom are they seen.   To find a young red-leg is as difficult as to find a four-leaved clover.

And when the marshes are alive with the voices of all the watery tribe of birds, you may one sweet evening in May "happen upon" one of the most beautiful and poetical sights ever witnessed by man—their exodus to the waterside.

So soon as the egg-hoods shall have been cast off, even if the young lie trembling in their nest a long mile from the water, with intervening morasses, dark forests of chate and reed, broad rivers, still waters—all these vast obstacles quail not the hearts of these wild little spirits, and directly the last-sprung shell is pulled off, the parents begin hovering in the blue over their offspring, calling *tëa, tëa, tëa*, and the pretty little heads look up affectionately and trustfully, and the great voyage across the dark continent begins, the parents flying about a mill-length over them, and a yard or so in front, calling anxiously *tëa, tëa*, and the brave little souls plunge, with chirpings, into a soft morass, and half-swimming, half-wading, they struggle through the nauseous mud and marshy mephitic air, dodging round the big stumps of the reeds—on, on, following the shrill *tëa, tëa*—on through a dark jungle of reed, where lurks the huge stoat, whose yellow head bursts like a tiger from the jungle and seizes a poor agonised little creature, bearing him off into the gloomy recesses of the reed-bed.   Scattered by fright, the remaining children run wildly hither and thither through the dark reed-stalks, or crouch low whilst the agonised parents call with heartrending cries from over the reed-bed forest.   Running and breathless, they arrive at last upon the open marsh, having left the gloomy reed-forest behind them, and one poor little brother.

But the smiling marshland, gay with cotton-grass and ragged-robin, entices them, and they make a safe and pleasant passage across the grass, having crouched low once as a gloomy harrier flew lazily overhead, his dark shadow filling their souls with dread ; but their brave parents attacked him, and drove him off, like a great bully.

Tëa, tëa, and they stand upon the bank of a broad river—their first dike ; and though 'tis their first essay, they plunge right boldly in, and are soon sailing gaily across, following their clear guiding *tëa, tëa*, when suddenly a loud squeal startles them ; a huge and horrible snout, with great sharp rows of teeth, shows for a moment where the baby was, and a sister has disappeared below the water, swallowed by a hungry pike.

Amid shrieks of despair, the old birds sweep down and cry anxiously to their young children to hasten across the dangerous river; and when they climb exhausted on the opposite shore, having lost two little sisters, the cheery *tëa-tëa* leads them on through a coppice of chate, with its sharp-edged leaves, on through forests of rush—for they are nearing the broad—until at length they halt, exhausted, staring over the inland sea; and the anxious parents fly down beside them and begin feeding them, as they gaze between each mouthful at the far-spreading, glistening waves, where the strange grasses float idly. And in some cases they adventure this inland sea, the old birds flying about and alighting on the sluggish waters, calling them, steering them across the great silent lagoon, as surely as any compass steers the mariner.

It is a strange migration, full of sublime terror to the young birds; and yet a commonplace incident in early summer, and more readily to be seen than the young snipe's exodus, as red-legs are noisier, the snipe going down to the water in a quieter, more stealthy manner.

Red-legs seldom travel farther than a quarter of a mile in a day, but always in a straight line for the water, even passing through gardens, across roads, under hedges; like the lemming, their course never varies by a fraction of a point.

Nor are the birds exempt from dangers when they grow

old in the nesting season, for the season is long, three nests being built if the early eggs be robbed.

I will tell you of a tragedy witnessed by a reliable gunner,[*] who was beating the marshes in search of eggs one July.

The day had been hot, the heavy air full of midges that stung, raising great lumps on the faces of the unseasoned, when the sky to the westward turned black as ink, and a few flashes of lightning played vividly across the inky pall— the landscape looked dull and lowering, as if the deluge was coming, for heavy rain had begun to fall, though the sun was shining brightly on the marsh at his feet—when suddenly a glorious rainbow burst forth, its many-coloured streamers rising from a distant group of trees, and a sharp wind squall rent the reed-beds, making the sail in a wherry volley like thunder as it shook its thick folds. After the "roger" and "low" had passed, a bright after-glow lit the scene, and the fenman went harrying the marshes for eggs, passing on his journey a clump of long gladen shooves drawn on the bank, and resembling the bodies of drowned maidens with drooping, lifeless heads, when he suddenly stopped. An old "buzzard," as they call the marsh harrier, was beating the marsh, busy hunting for eggs and young birds. As the bird drew near, he saw it suddenly sweep down into the marsh-stuff, where it was lost to view. Taking his mark, hunter-like, he skirted a reed-bed, and suddenly came upon the gloomy buzzard eating a hen red-shank, and there close by were her eggs, hot and shiny.

The harrier had seized her from her nest, and the cries of distress of the sentinel cock were accounted for. Sometimes, too, you will come upon a "lone bird," or widow, with no sentinel; he has been killed by a smart gunner or prowling bird or beast of prey. Still she does her duty, and rears the young brood unaided.

---

[*] One gunner followed a family migrating from a marsh on the *sea*-side of Waxham Church to Hickling Broad, and in a turnip-field he tried hard to find the young, but could not see one.

When the guns **begin to boom in August,** the birds flock; and though the young **birds** are nearly as large as the old parents, you shall know them by their lighter-coloured coats. At that season, too, you may come upon a belated cock sitting upon a post, calling to his young, who are feeding in the **marsh** crops by the waterside. All day and all night **you may see the noisy sentinel on a post, calling to the young, and nodding its head.**

At this season marksmen try to shoot them, **but only** the **most** experienced gunners kill them, for they cannot **be** killed when hanging, as would easily appear; they must **be** fired into, when they turn away. It is said **amongst the** gunners they see the flash when hanging, and **have ample** time to turn away from it.

Just before harvest they leave the land of their birth, and seek the ever-broken and restless waters. Down there on Breydon flats you may see them on a steaming hot day, when the landscape hangs quivering in the air, and the tide is softly creeping over the flats, playing together and feeding—dancing in the rising silvery flood, flying **off** this **way** and **that in zigzags,** their breasts buried in **the** water **as they swim beautiful** figures over the tide, their **wings** hanging **loosely by their** sides—some looking gravely **on, others** bounding with fun **and** merriment. But the **rising waters** drive them to the beach, and ere the September moon **is full** they have joined the other wandering tribes of birds and gone to other lands across the grey sea *—and have left the fens to the winter and to me.

---

* In Anglesea they winter all **the year.** In the winter of 1891–92 I saw red-shanks every week.

CHAPTER XCII

CURLEWS

THE old "calloo" is a frequent visitor to the Broadland, in bunches of a dozen, two dozen, or three dozen. At all seasons of the year you may hear the *cur-l-ooë*, *cur-l-ooë*, *cur-l-ooë* of the leader, as they fly from the wet sea-beach to the marshlands or uplands; and the crafty gunner often lures them to fly over the reed-bed, where he lies hidden, by merely imitating the leader's whistle, or his wife's, for they have different whistles. You may lure them from afar by this simple trick.

They are to be seen round about the Broadland any month of the year, feeding by the ebbing tide, seeking grubs and snails on the moist marshes, hunting for worms in the wheat-pieces, delving in the dried cakes of cow-dung for *larvæ* on the hard grass marshes; for it does not take much food to keep them alive, as it does a snipe or woodcock.

They are mostly shot at the morning and evening flighting hours, as they fly low on these journeys to and from the uplands—the gunners whistling them down, even turning them in their flight, and bringing them right down close to them; and when they are shot, their black-fleshed breasts are eaten—a dish I care not for, though highly esteemed by some.

One gunner told me one once came straight to him to see what he was, and he knocked him over with a stick. Inexperienced gunners mistake the snipe's flighting-note, *tuttuo*, for the similar but harsher curlew's flighting-note.

But curlews are mere visions in the Broadland, coming and

going, flying in formless battalions to the green uplands and grass marshes, the feeding-places they love best, for often in softer feeding-grounds they have to bury their scimitar-like bills to the hilt to catch their prey. And though they are to be seen throughout the summer, they never breed in the Broadland; indeed, they must defer breeding for the first few seasons of their lives.

The Whimbrel, or "Half-Calloo,"

In habits, custom, and appearance much resembles the curlew. He too is to be seen in smaller numbers thereabouts; he too loves a grass marsh or a grassy upland; he too flights and whistles as he flies; he too does not breed. Nor have I seen him in the winter months in this district. He too comes in increased numbers in August, the young birds from the nest having wandered down the coast; and he too is not much to eat, but merely gives us vague thoughts as he or his larger relative flies high through the closing day, whistling with melancholy voice over the wide, bare marshland.

CHAPTER XCIII

THE COMMON "TERNER"

WITH the month of May come the "dahrs," and on a fine
day, when the sun gleams from the wavelets raised by the
gentle sea-breeze, you may sometimes see a number of these
graceful birds playing about over the water—hawking for
flies, which they chase and catch after the manner of a
puit or night-jar; indeed, in the summer dawn 'tis hard
to distinguish which of the three birds is hawking over the
reed-beds for the "mingen," that have got up early to enjoy
their ephemeral lives—the birds climbing the air, turning this
way and that, seizing the "mingen" by the hundred.

And with May the terns are gone; though I saw one a year
or two back in August, hawking at dawn for flies above a
reed-bed, filling its crop, as I suspect, to take home to some
young ones not far away; but no nest have I ever seen
taken, though one reads of colonies of "blue dahrs" in
days gone by.

HERRING-GULLS.

CHAPTER XCIV

THE BROWN-HEADED GULL

So certain as March, the roarer, comes round, bringing dry winds that rattle the dead reed-stalks, and raise clouds of dust along the long straight roads of the marshlands, so surely do the puits return to their islets on a certain broad: their last abiding-place. For though they have in recent years essayed to form a colony upon Hickling, where the first nest was robbed and the second pulled out, they try no more, assured of its futility. Nor do they return to Somerton Broad, where I saw a colony as recently as the summer of 1885. They dare not; for their eggs were gathered and sent to market by the bushel in the brave days of old; and though they deserted Somerton Broad some years previous to 1885, they returned thither once more to see if the natives had learned charity. But a provincial population never learns that virtue. And the birds know it, for they go only to the bound and chained waters of Hoveton, where, punctually—to a day, the fenmen say—the first stragglers arrive from across the sea, coming inland in a V-formation, their crests still pale white, for the brown cap grows a little later. And at this season, when the reed-cutters are still busy harvesting the swampy crops, the wild, strange cry—a voice like the sea—of the puits fills your heart with gladness, for you know the spring-time is come.

And you have not long to wait for the rest of the colony; day after day their numbers increase, and you see them flashing athwart the blue sky with wild cries and settling down.

And if you watch the colony, and can bear with their wild cries, that can be heard for miles, you will see strange flutterings and bustlings on the shore—the pairings—and you scarce can tell when they have paired. When April shall have run her round, comes the serious business of laying the blotched eggs, in their rough nests of broken-down reed and gladen, slim and precarious lake-dwellings. These nests are often placed closely together, two nests occupying a space as large as my writing-desk. And there is such a hurrying and commotion all through incubation; and when the beautiful little downy creatures leave the nest, as they do a few days after being hatched, all the broad is a wild scene, recalling a Norwegian fiord. The little puits swim around the little lily leaves, eating flies and worms, whilst the old birds dart about, gleaming against the azure, and filling the welkin with their mysterious oceanic voices. And should you appear, pushing stealthily over the waters from behind a reed-bed, the nearest birds will topple and throw themselves over in the vernal blue, shrieking and fluttering down, trying to hide their young, who swim quickly to the reed or gladen beds, for I have never seen either fledgling or old bird dive.

Some birds, having been robbed, will be seen sitting on a second clutch; and perhaps you may see a great saddleback gull suddenly sail out of the sky and sweep down amid a noisy commotion, killing some young with bill and claw, amid the furious assaults of the white-breasted mariners. But these pirates are always successful, as are the rats, who steal at dead of night upon the eggs, devouring them; for your rat is a deadly enemy to the puit, and many another bird besides.

And when the young mouths are many, you see the old birds hawking till dark, when they perch (rather clumsily) upon the broken-down gladen-stalks, now a formless mass of decaying vegetation; for they delight sitting upon stalks just rising above the stagnant waters of the lagoon, and watching for the first flies that hover over and dimple the silvery plains, for the puit is, like most migrants, an "insect" feeder.

And at this season, when the hurricanes are raising the dust in clouds, and the foul grass smoulders and smokes into the blue, and the patient horses pile the hard clods on the broad lowland, the puits follow the strong plough, in company with starlings and rooks, feeding upon the insects, and worms, and snails turned to the surface. And a picturesque sight it is to see the gleaming puits and sable rooks following, with the crested lapwings and restless starlings, against the sandy dunes, where the marram shivers in the cold nor'-easter.

And these are the "gulls" that follow the plough—for not another kind have I seen following the ploughman. At this season, too, you may see them on marshes bright with water, feeding in a large flock with peewits, and then you know the peewit is the sharper-witted; he is more alert, takes alarm first, and flies straight up in the air, disappearing in the blue, whilst the hungry puit merely flies across into a neighbouring marsh.

Far a-marsh they go from "their broad" in their journeys for food, and at eventide you may see them flying back across the chequered landscape—a beautiful vision of white flitting bodies against the dark April blue—when the night-jar is filling his crop with food to carry to the young ones at home.

I have watched him soon after dawn hawk for a full hour —*sans* intervals—over a reed-bed, darting down quickly, rising gracefully in the air, turning this way and that, behaving like a tern or night-jar; in fact, I have known him mistaken for both these birds in a dim light. And when the young can fly and shout, away go the colony to the wet sea-beaches, and to the mud-flats, to feed on worms and sandhoppers, and other insects; and for a short period you may hear their voices in July by the ever-restless sea, ere they return over the grey plains whence they came.

And the mere sleeps tranquilly again; and the farmers around have lost a firm friend in that restless, graceful, embodied spirit of the sea—the brown-headed gull.

NEST AND EGGS OF THE BROWN-HEADED GULL. (*In situ.*)

CHAPTER XCV

SEA-GULLS

ARE constant visitors to the Broads. They come in "ter wash their nets," as the natives say. And, in truth, they do come in (when not storm-driven) to wash, and take in fresh-water. At all seasons of the year you may see their white wings flashing against the blue, as they sail over the sand-dunes, and drop with loud wild voices on to the still waters, where they sport, drink, and preen their feathers, whilst a watchman sits on duty; and should you disturb them, they will rise with their wild *laka, laka, laka*, and the *cou-l-ooë, cou-l-ouë* of the herring-gull, all flying away to sea to rob the fishing-nets, if it be the herring-season. And yet the sentimentalist raises his voice in the land, regardless of these depredations on the fisherman, and eke to the landowner; for the great "saddle-backs" are as ruthless in their search for eggs or young birds as are rooks or crows. I have seen them beating in early spring over the land, flying low, on the look-out for eggs or tender fledglings.

In September, when the reeds are yellow and the waters blue, the common gull and the herring-gulls come to the broads in increased numbers, and hundreds of them may be seen idly floating upon the still lagoons—into which they often "scorf up" whole herrings—dreaming, mayhap, of the sea, whose roar comes over the dunes, while others come and go, or settle with them; and on a bright sun-shiny day hundreds can be seen sitting motionless on the waters; or, if the winter be hard, and the land be white, and the broads be laid, you may see them sluggishly resting on the bosom of some salt æstuarial water—as Lake Lothing —the cold and biting snow and hail beating down on their glossy backs.   But what would you ?   'Tis better than being upon the stormy grey sea.

The saddle-back gulls, too, great and small, are to be seen often in the spring and summer; nor do they stop long on the waters, merely coming in to wash and drink, and do a little poaching, ere they return to the sea—their home.

None of these birds are to be seen for long on the broad, nor have I seen any of them following the plough—as poets have written.   The " gull " that follows the plough is the " puit," or " puit-gull," * for they love worms as dearly as a gull loves a shrimp.

But ever welcome are the strong bright gulls, with their flashing bodies and wild sea-cries — birds cradled upon the blue sea, upon whose restless bosom they must seek their living or perish—birds into whose bodies the spirits of the Norfolk sailors used to go—fishermen's brothers, since their old men are " tarned ter gulls "—for ever doomed to rove the restless deep—brave, and hardy, and free !

* Not " pewit-gull," as Mr. Saunders writes it.

HERRING-GULL'S NEST AND EGGS.   (*Yorkshire.*)

CHAPTER XCVI

GREAT NORTHERN DIVER

WHEN the March winds rustle the frayed, sere, amber reed-stalks surrounding the blue waters of the broad, and each particular reed stands bright and clear, and the lifeless tree-forms bite sharply as an etching into the hard blue sky, and the landscape is hard and metallic, the fine great northern diver makes his appearance upon the cold waters, for then the fish are in the finest condition for man and bird. At this hard, cold season of the year, though the spring equinoxes hurl the cold sea-water upon the wet sea-beaches, and the stars glitter hungrily by night, and the winds howl, still, above the noise of the storm, is heard the melancholy howl of the great northern diver—a wild voice of defiance to the wilder voice of March. And by day, if you watch him, you will see him fishing for silvery roach and red-finned rudd, diving into the hard blue waters, and taking fish several inches in length, as you may prove by shooting him and holding his spotted, heavy body before you, and shaking him, when the fish will come streaming from his bill—rudd and roach. I have known sixteen counted from one bird's stomach.

But the wariest of gunners will tell how difficult he is to shoot—that is, after he have "been over" a little, and got used to the artful ways of the craft. When he first arrives on these inland waters he is tamer and more confiding, and all he asks is to be let alone—to poach. And his brother poacher knows this, but he will none of him, so he shoots at him upon the first fair opportunity; but should his "old Emily" miss, and it sometimes does, the great bird is not going to give

him a second chance, if he can help it.   He will dive under
the blue with eagerness, and go forty yards under water at
a rapid stroke, for he is a strong paddler, as strong as the
punt-gunner who is after him.   Indeed, old gunners say he
is the strongest bird, with his feet, afloat.   Up he comes,
sparkling like a water-flower in the sun.   But he will not
rise on the wing—he knows too much for that—and away
goes the excited gunner sculling after him; but down he
goes again, and so on for hours he will carry on the game
of hide and seek, for he does not fear the cold.   And yet
the weakened gunner follows, his heart full of hope, for he
knows whichever way the bird puts his head when he dives,
that is his course.   Shot after shot is wasted—the shot
merely dimpling the hard blue waters of the lagoon—until
at last, tired, perchance, of the chase, out of mere sport the
creature takes to flight, and escapes from the tired and excited
gunner, who trembles as he fires his old rusty fowling-piece,
and the harmless shot whizzes through the air, falling like
lead upon the lagoon further ahead, and the big speckled bird,
with a groan, flies over the dunes out to sea, where gunners
are not.

But the most artful gunners work differently ; and one who
has shot many tells me he can always make sure of "my
gentleman" by going down to leeward—before the wind—
and suddenly startling him, when he will probably take to
the wing, and then he falls an easy prey to the old ten-bore,
and is taken home to make his wife a muff.

But this wild sea-viking does not stay with us long—the
softer air of April does not seem to suit his hard, savage nature
—and he rarely comes in from his home, the sea, until fierce
March returns again on the revolving wheel of the seasons,
and again the gunners are alert to muff him ; and so the
sport is kept up—dog robbing dog, poacher killing poacher.

CHAPTER XCVII

GREAT CRESTED GREBE

THESE proud domineering birds are not uncommon on the lagoons, where you may at all seasons—except in the depths of the hardest winters—hear their loud clanging voices. In the mild days, common at the end of February, you can see them begin courting, chasing each other in and out of the brown gladen or amber reed, splashing and diving after each other, at times rising and flying down the lagoon. At all hours of the day and night I have heard their strange shouting voices. If the day be fine, you can watch them courting near by, for they are not very shy. Then you may see the male and female swimming round each other, nodding their heads, setting out their wings, arching their backs and finally diving, one this way, another that, coming up yards away from each. Some old Broadsmen assure me they tread under water; others, that they tread standing straight up, breast to breast, embracing each other with half-outstretched wings, as the old coots also tread. Or, again, you may see two chasing a hen, pecking each other until one retires from the contest. In the reeds, too, you may hear them growling and splashing, but I have never seen a real set-to fight.

At this season the cock-bird's voice is strange; for he growls a challenge, recalling the mixed voices of goose, cock, and dog—a harsh voice it is too. At other times, when swimming in the open, he makes a noise best imitated by putting the tongue into the palate and making a quickly-repeated chucking noise with open lips—the sort of noise made in urging on horses. Then follows a coot-like cry (only sharper in timbre), followed by a hoarse *ah-ahwah-ah! ahwah!* followed again by the chucking note repeated some score of times, followed lastly by a hoarser *guo, guo-oh!*

Their nest, made in the middle of May, is merely a heap of broken-down gladen-bed, and resembles a swan's nest. They merely break the sere gladen down, making a pile from four to twelve inches high, which floats on the water, rising and falling on tidal water with the ebb and flow of the tide-stream. On this pile of gladen they place weeds pulled up from the bottom. In the middle of the nest is a cup-shaped depression, in which the five or six eggs, the colour of faintest green chalk, and about the size of a silver-spangled bantam's eggs, are placed. At other times the nest is placed in a reed-bed, and is almost entirely made of weed—chiefly hair-weed—which they pull up from the bottom of the broad, making a great splashing and shouting as they build it, for they are noisy workmen. Sometimes the eggs lie in the water itself, and nearly always they get discoloured by the wet weed in the nest, or with the weed or rotten, sodden gladen, with which they conceal the eggs when they leave them, though at times they merely use gladen to conceal their treasures. One old Broadsman, whom I have ever found trustworthy, tells me they often lay first six eggs (and never more than six) and cover them over with weed, laying six more on the top of these. Then, he avers, they hatch off the top six, one bird taking charge of the nest-lings, whilst the other "gets up" the first six eggs, and hatches them off.

It takes three weeks to hatch off the beautiful nestlings, and thankful must the hen be, for she does all the sitting; indeed, the cock will drive her back to her nest if she leave it—that is, if he can; but she generally manages to get away for half-an-hour at a time to feed, whilst he sails proudly near the nest, driving off any bird that approaches his eggs, as does a cock-swan. If the hen get away from the nest, which she does by swimming first and flying off some two hundred yards afterwards, he will go at times to meet her, the pair swimming back to the floating hover.

When the young are hatched off, the family take to the water, and you may see them feeding on flies from the top of the water, or you may hear a great noise and splashing in a quiet reed-bush, and come upon them feeding the youngsters on small eels and fish—taken head first. At other times you may see the youngsters sitting on the old bird's back, after the manner of swans; and should you disturb them, the youngsters will steal off her back, some this way, some that, diving out of sight. They carry their young long distances round the broad, or up dikes, where they can catch their fish more easily. One old gunner once took eleven small roach from a dead grebe. If chased, they will fly under water like waterhens. Yet with all their precautions, they seldom raise more than three birds out of a clutch; for the young are not difficult to catch in shallow water, the Broadsmen sometimes capturing them in "fleet broads" by aid of their quants.

They grow large and are almost full grown before they can fly, for they are big eaters—flies, small eels, fish, and freshwater "clams" (mussels) being their staple foods. But in hard weather, when the broads are frozen, they will eat grass from the shores, and even corn, for they do not leave the broads in winter, even when the broads are laid—that is, unless every bit of water gets frozen hard as iron, when they go away, returning as soon as the ice breaks up. When the broads are laid, they will sometimes "huddle up" on the ice, hoping for better days.

Liggermen detest them; for they will clear their liggers of fish, never getting caught—"the artful owd varmin."

Indeed, the old Broadsmen say, "You might as well try to catch a fish with your hand as ter catch them."

But for ornamental waters I know of no more desirable bird; his beauty, personal demeanour, and wild clangour endear him to all true lovers of wild life.

GREAT CRESTED GREBE'S NEST AND EGGS. (*In situ.*)

CHAPTER XCVIII

THE "DOB-CHICKEN"

STRANGE as it may appear at first sight, in all that vast acre-
age of broad water and river that roams and dots the Broadland, but few "dob-chickens" will be found; 'tis really a
rather rare bird in the district, and chiefly to be met with in
the quieter rivers—not that it prefers running water everywhere, for I have seen numbers disporting on a great Irish
loch (Lough Erne), but never far from the shore.   However, in the Broadland he is rare; and methinks he cares
not for still and brackish water, and that is the secret of his
scarcity; for most of the rivers (below locks), and many of
the broads, are brackish, in some places seventy-five grains
of "salts" being found to the gallon of water.   Soakage
from the sea, salt springs, tides, and saline water pumped from
the marsh dikes, which in the winter hold salter water than
outside, all contribute to the salinity; and I have a notion
that such waters are not beloved of the grebes, for the great
crested bird loves best the inland broads, such as Ormesby,
Wroxham; and the lesser grebe, or dob-chicken, I have
seen in greater numbers about the locks on the rivers than
elsewhere, though I have seen them about the outlets of the
mills, where the water is shallow, and often running purer after
rain; perhaps he can catch his fish better in running water
—at anyrate, I have never seen any on the open broad, or
upon the open lakes of Scotland or Ireland, but always round
the gladen and reeds.   But never in all my wanderings on
the broads have I seen the nest, though I have met a few
"eggers" who have at intervals found their nests, which are,

as is well known, like those of the great crested grebe, but on a smaller scale. One old egger, who has found three or four nests in this district, tells me "they were all laid in turf-decks in fleet water, and every nest had six eggs, and the old bird cover her eggs up when she leave them."

In early spring you may hear them calling each other through the February mists with their soft whistling notes, and if you be wise and attend the signal, you may see them swimming near the shore, working up stream. I think they sight their food before they dive, they seem to make their dives so suddenly, giving a quick dart forward of their heads when they go down, as if striking at something visible. At a distance they appear to go under water at each dive.

SEA-GULLS AND NESTS.

PART II.—BEASTS AND FISHES

(MAMMALS, FISH, AND REPTILES)

SUNSET ON SALHOUSE BROAD.

" They hung the slaughter'd fish like swords
On saplings slender,—like scimitars
Bright, and ruddied from now dead wars,
Blaz'd in the light,—the scaly hordes."
—T. CRAUFORD, *The Canoe.*

FISHING ON HOVETON BROAD.

CHAPTER I

BATS

TWO bats are common in the Broad district—the common bat and the large bat, measuring fourteen inches across the wings.

THE COMMON BAT

may be seen any day of the year, but he is very uncertain, and his appearance depends much upon the weather. On mild winter evenings he is sure to appear whenever there are any midges flying about, and you see him hawking through them, flying to and fro thousands of times; and you can hear his jaws snap, the sharp little mole-like teeth coming together with a click. Like the tom-breeze (dragonfly), he hawks to and fro, devouring thousands of midges. But he is never to be seen in coarse weather: he lies up then in an old woodpecker's hole in a hollow tree, or in some safe retreat.

The bats' appearance is at all times very uncertain, but just before dusk is their hour; and if they appear in winter, the Broadsmen say "some wild weather is a-coming."

THE LARGE BAT,

common to the district, I have seldom seen in mild winters even; but in the summer he is common enough, appearing over the dikes and rivers from half-past seven to half-past eight or nine; he and the night-jar often turn out together, just as the cuckoo is going to bed.

He has a lustful, doggy face, and a yellowish, ferret-like fur. His wings are large and strong, and he is very tenacious of life. He flies round shrieking, and sweeps down on his prey like a swallow ; and like a swallow he captures flies from the surface of the water, dipping right into it, rippling the still dike or broad. He hates light; and should you take him into a lighted room, he will shun the light and sneak into the darkest corner. During the day he hangs in some leafy tree.

I have opened several of their stomachs, but never found anything but flies and midges—never a beetle.

CHAPTER II

HARE

THE hare is familiarly called "old Aunt," or "old Sally," by the marshmen, for they look upon him as their legitimate property—as an old pal who will often come in handy—in exchange for a few shillings.

In March (or earlier, if the weather be open), when the reeds are yellow and the rush dry and sere, you may come across the bucks fighting on a moonlit night in a rush-marsh, standing on their hind-legs, "smacking each other in the face," and felt-pulling. Next day you will see large forms in the rush-hillocks, with fur scattered about, and you know that other fights have taken place. This is the solstice of the love season ; for though young leverets may be found every month in the year, March is the love-making month—the month when the bucks (who are a bit smaller than the does) will go miles of a night, running along mad with mere desire (and then your experienced nooser sets his wires low) on the doe's scent. Perhaps five or six bucks will start after one doe, and all pick her up, when there is great commotion, the bucks jumping about, running into each other, and jumping over each other, and fighting all for the doe. At last one gets her ; and so exhausting is his love-spell, that he often falls off backwards after the act, perfectly dazed, as in a fit. At this season they are restless at the top of the day too, and may be seen running about sniffing the ground after some doe or other ; and you may tell them as they run, for the buck runs with ears erect, whilst the doe runs with ears laid back. She is timid, mayhap. Indeed this is the season to distinguish easily between a buck and a doe; but experts can go wrong, though they can tell rightly four out of five times. And hares prefer at this season the rush marshes, the thickly cropped wastes where the rushes grow in scattered islets in a sea of soft, withered white grass—marshes not beloved of birds, though the grasshopper-warbler is at times found there. In these rush-marshes, in great rush-tussocks, you will find the moist and cool forms marked with their " feetings," and littered with wool, signs of recent fights or amatory struggles.

As they " go " for twenty-eight days, like a rabbit, you may find the young in April, and indeed they litter every six weeks throughout the year in greater or lesser numbers. But April is the month when most young leverets are dropped upon the new-lays, in ploughed fields, or on the marshes—in short, everywhere and anywhere. They generally cast two or three at a time, but four have been cast at once. When the doe casts her young (who cannot see for nine days ?), she leaves them at once, sometimes returning to suckle the *open-eyed* (?), sluggish creatures for three weeks, when she leaves them to their own resources. Sometimes she deserts them as soon as they are born, when they die. Perhaps such an one is more anxious to begin breeding again than most of her tribe, for they all begin again a fortnight after they have cast their young. The sluggish young do not wander far from their birthplace until they grow to three-quarters their full size, trusting to their sight to dodge an enemy, and to their colour, which protects them in a marvellous manner. They are difficult to see even when pointed out—that is, until you get accustomed to such sights. Should you find young leverets, and wish to see the mother, put the back of your hand to your opened lips, and make a sucking noise (such as you can decoy rabbits from their hole with), and you'll soon see the old doe come up amongst the rushes, staring at you curiously; but do you make that noise in winter, and they are " going." Some of

the marshmen call hares to them, as some do partridges and
pheasants (by aid of a bit of sand-paper). By this method,
on moonlit nights, the poachers make this sucking noise if
they see an old hare, and it generally sits up on hearing the
sound, and gets knocked over. When the height of the
love-season is over, they get more regular in their habits,
and return to their forms (often near the villages) at the peep
of day (which is the best time to shoot a hare), and lie there
with their hind-legs beneath them, and their heads resting on
their fore-paws, lying as a cat lies, but with one eye open;
for they see any one directly they enter the field, though
they do not move. They like a sunny place for their forms,
and sleep most in the morning, being more wide-awake
after two o'clock in the afternoon. They will return to the
same form for a month together, but should they be hunted,
they'll return to a favourite form on the third day; but often
they will make another form close by the old one, sometimes
within a yard of it. They do not get shy of villages or rail-
ways either. I know a field through which the line runs
where I have seen fifteen hares turned off in an afternoon.

At "shutting in time" they draw out for food, wandering
miles, even swimming broad dikes in search of it. They eat
grass, young clover, young rushes (which they bite through),
turnips, wheat, mangolds (not a favourite food), carrots, celery,
cabbages, vetches, rye, young poppies, sow-thistles, dande-
lions, and even hay.

When you study this list of foods, when you think of those
long greyhound-like quarters of theirs, and the speed with
which they get over the ground, and when you couple these
facts with their wandering propensities, you begin to wonder
what harm they cannot do to the farmer. A hare will nibble
a hundred wheat-straws in a night; he will bite into endless
roots, leaving depressions where frost and water lodge and
foster rot; he will tear the hearts from the clover crops;
and when you know seventy hares have been counted in a
field, you begin to wonder what the farmer loses by these

creatures; for though it is well known that the biggest enemy
to wheat is wheat (when it grows too thickly), still the hare
is a pretty formidable enemy, and when the corn is up, you
may look in vain on the marshes for hares—they all get into
the corn.   Even the keeper of a good game-preserve does
not grudge the hares being killed, for he can't look after
them—an he would.

You can easily decoy hares from one marsh to another by
placing carrots about ; but there is no surety they will stay
there. They are uncertain and wandering creatures.   But
one thing is certain, that you might as well turn a flock of
sheep into some fields as a flock of hares.

When harvest is over, the hares draw to the covers and
plantings round the broads; but they begin to leave them
as soon as the leaves fall, and draw back to the marshes to
their cosy forms in the rush tussocks, where you may find
them at the beginning of winter.   And it may be remarked
here that they do not scrape a hole in their tussocky haunts
as they do on a field, for there is no need to crouch flat with
the earth when you have tall rushes growing over you.
When they take to the marshes again, their habits become
more regular.   They have regular routes they frequent,
often going long distances, and stopping out a day or two
in some temporary form *en route*.   In these highways the
" looking-glasses " are set by the needy marshmen.

When the marshland becomes covered with snow, and the
dikes are laid, they may be found in snowy forms—really
snow-houses.   Should you disturb them, they will gallop
away, and try to throw you off the scent by making long
side leaps before they squat down.   And in this manner
they may easily take you in, and leave you thinking them
lost.

At this season they often get drowned, for they will
start to walk a frozen dike, the thin glaze gives way, and
they take to swimming, breaking the ice as they go.   Sometimes they get across safely, but at other times they grow

exhausted, and give up and are drowned. They never seem to think of turning back, though one would imagine creatures so fond of the water as they would be more at home under the circumstances; for you may find them sitting in the water summer and winter, and often on the marshlands they plunge straight into the nearest dike if you surprise them. Also, they take to the sea if pressed, as also to the broads and rivers when the harriers are close upon them. Old gunners have told me a dog is sometimes drowned in the ice in the same manner as the hare, a bitch never—but I know nothing of it.

A favourite resort for hares in hard winters is the reed-bed. In the hardest weather, they draw up to the garden, and eat the brussel-sprouts. I know one cottager who killed nine in one winter in this way—three of them at one shot. They came across the ice (which was sixteen inches thick) to his garden, and he was avenged.

For coursing, the marshes are "the place;" for when the hare gets to cover, he is safe. When hunted, they run in circles, coming up in the field whence they started should they not be wooled; but if the hounds wool them, they'll run straight away—bolt wildly away. Nor are they sure to take to the "wall" again, as they are when disturbed on a marsh, as the knowing gunner knows well.

CHAPTER III

MICE AND VOLES

THE marshmen distinguish four "mice" and one "water-rat," three of which are really voles, viz. :—

1. House-mouse.
2. Field-mouse.
3. Red or "sondy" mouse.
4. Marsh-mouse.
5. Water-rat.

} All really voles.

1. THE HOUSE-MOUSE.

The house-mouse is too well known to need any description.

2. THE FIELD-MOUSE.

The field-mouse has large eyes and ears, and is long-tailed. He breeds in the fields, hedgerows, heaps of manure, and you may find his young (seven or eight in number) any time from April to Michaelmas. He is omnivorous, but not ubiquitous, for he seldom frequents the marshes unless they be cultivated. They serve as food for cats, weasels, stoats, owls, and kestrels, though all of these prefer the "marsh-mouse."

They say in Norfolk, "If you see a field-mouse up in the stuff, you may be sure of a wet harvest."

3. THE RED MOUSE.

The red mouse is found mostly in the autumn, at thresh-
ing time, frequenting fields. They build their pretty little
nests in the grass, in the harvest-field, in hedgerows, and
in corn-stalks, but chiefly on the land.

4. THE MARSH-MOUSE

Is a very short, hairy-tailed, pretty little creature, that
seems to be born to feed kestrels, hawks, and weasels, for
these innocent creatures are the staple dish of those marsh-
rangers. The "mesh-mouse," as the outside men call him,
has a beautiful brown coat and whitish belly, his head and
tail being shaped like a water-vole's, his ears being rather
large for his size, but they are hidden somewhat in his fur.

And should you wish to catch one to study, you had better
follow the marshman's custom and mow a "clear" as large as
your writing-table; then go some yards to windward of the
"clear," and start mowing, going before the wind towards
the "clear." In this way you will frighten the marsh-mice
into the "clear," when you or your mate may catch them.
The marsh-mouse, which is the commonest of all the "mice"
in the marshland, is a vole, and vegetable feeder, eating lily
roots and grass.

Their nests are found on marsh bottoms from the latter
part of March to October. These nests, made of fine soft
grass, are placed in depressions on the marsh, and have
round holes at the sides for doorways. And as they are
most numerous in June and July, the marsh-mowers turn
up hundreds of them.

Marsh-mice can swim; and as the water-vole swims
faster than a rat, so the marsh-vole swims faster than a
field-mouse. Pike and herons sometimes take them on
their journeys to and fro "athwart the deeks."

5. THE WATER-VOLE.

When the warm spring breezes ripple the glassy dikes and wide lagoons, the beautiful chestnut-coated water-voles leave their dark homes in the pretty dike-sides, and go in search of food, for their winter store has got very low.

You may see them eating the bark from the sallow bushes, nibbling the chate and reed roots, feeding on young rushes, or else biting into the fenman's turnips, an they should have survived the winter snows and ices. In old reed-beds, too, you will come across them feeding upon old reed, or else you may see them carrying bits of these foods to their holes just above high-water mark, the nibbled bits of green stuff often betraying them.

All day long, through spring, summer, and autumn, you may in secluded dike-ways come upon them playing in parties, or perched on the banks washing their faces, or plunging into the stream and swimming hurriedly along the shores to their holes—their hairy bodies and tails threading patterns over the dike-ways, where the succulent duck-weed pies the water-ways with green leaflets. At times you may get a good peep at them, for they are impudent little fellows in spring, and will often sit peeping from their holes with upraised heads; or you may come across them on the marshland, when they will squat and trust to your overlooking them, instead of trusting to their heels.

Should you dig a burrow out, you will find it goes straight in for about a foot, then turns upwards towards the right or left, the nest being a cavity large enough to hold a peck, and securely placed beyond the reach of any floods. In this hollow they make their nests in April, using warm dry grass and chate. In these nests the young are born, but these I have never seen; though I knew an old fenman who told me "he'd seen the old one carrying the young acrost a deck like a dorg carry its young."

If, during the summer, you come across them at close quarters, and "stow them up," they'll dive, and coming up again at a distance, put their noses just above water, as does a coot, and eye your every movement. During the summer and autumn they fill their store-houses with grain stuff, and most especially with horse-beans, when a patch grows near their burrow, as is often the case. Directly the first frosts come, and the reed-tassels and the money-spinner's webs are turned into jewels, they take to their houses, and do not appear again till the binding frosts have gone. On the mild days in winter, however, some are to be seen, but it seems doubtful whether they all venture forth from their cosy storehouses on such occasions.

But even they have their enemies. Hungry pikes in the dikes, and herons hard pressed, will eat them, whilst the fenmen shoot them as food for their fierce ferrets.

Nevertheless, enough of these harmless creatures survive to add an interest to the secluded dike-ways, which they most frequent, preferring them to the more busy tide-ways, and the "black" voles at times, moreover, wander as far as the farmers' stacks.

CHAPTER IV

THE MOLE

In February the moles begin to throw up their heaps on the sere marshlands. Those heaps vary in size from nice piles of earth that could be held in a child's toy pail to piles the "size of a tumbril-load." But the average measure of the nesting heaps is four feet in diameter. And though they work by night and day, it is a slow process making that intricately galleried nest; for they pat all the gallery walls with their flapper-like claws, leaving their nail marks upon the mud partitions. From this galleried house there are numerous runs—galleries running into the hard marsh-land—galleries bored through the marsh, as the smaller piles of "diggings" prove; but in soft upland earth, or in river-walls, they often merely force their way under the turf.

Let us, on a fine March morning, go down to the marsh, and dissect one of these domed and galleried nurseries. The roof-crust is hard as we cut into it, and we find—after having removed it—an upper circular gallery round the dome, with four descending galleries, leading downwards and inwards to the central hard-roofed globe, where the nest, made of fresh or light coloured grass, lies. Below this are galleries, leading downwards and outwards to a circular gallery under the turf—a gallery opening into several runs, that lead away into the marsh bottom. Every gallery is plastered and beaten hard by their little hands, the thousands of nail-marks showing the toil and trouble they had to build their home *with* hands. As this is a hard clay-bottomed marsh, their whole scheme of working is

laid bare.   The earth from their borings is generally heaped
at the end of a new run, a sign they had begun to work
vertically downwards, and perhaps, too, to allow themselves
space for turning round, just as, doubtless, the upper circular gallery of the cradle is for the purpose of allowing the
workmen to beat that extraordinary hard roof to the actual
nest—that upper arch of the central globe.   On the uplands,
on the other hand, they will work down eighteen inches in
depth, and turn up sandy gravel.   But in either case a fresh
nest has a crumbly-looking heap over it.

These earth-heaps over the nurseries also serve another
purpose—they are stocked with worms ; for the mother
suckles the young for a month or five weeks, until they are
old enough to work for themselves.   Should you approach
a nest when the young are large enough, they are sure to
hear you, and run off, and you find but the warm nest ;
for they are either very quick of hearing, or feel every
vibration.

Towards the end of April you may find young smoke-
coloured moles with unsightly pink snouts—three, four, or
five to a nest—and so on till the middle of May.   Then
there is a pause, and a second family is sometimes born in
August.

They work at all hours of the day and night.   There is
no surety when they will work, but generally after rain.
Superstition in the Broadland says they work only at
8 A.M., noon, and 4 P.M.   Others say the Jacks begin to
work between nine and ten o'clock ; others that females
work at eight and four ; others that both work all day
long ; and yet others that the Jacks leave work on fine
sunshiny days at 1 P.M., and "it bëant no use ter look arter
them when it hev come one by the day."   But all this "I
must leave," as they say in Norfolk.

They work the earth past themselves in the run ; but how
they raise these heaps is a mystery to me, especially the
large breeding heaps.   They will not "skim" (work just

beneath the soil) in dry weather, for the worms descend
for moisture, and they follow; but after a shower of rain,
when the worms come up, they will skim, and you can see
the worms coming out of the ground before them. Just as
when frosts are coming the worms know, and they bore
down, and the mole follows, working deeper; indeed, some
aged broadsmen regard the moles working downwards as
a sign of coming frost.

Old mole-catchers assure me the male makes a straighter
run than the female, and goes for a hundred or a couple of
hundred yards without turning to the right or left.

The young begin by skimming, and very superficial
skimming it is, for they are not strong enough to skim
deeply. They generally take their first lessons, too, after
a rain-storm; and you may, if quiet, see them working, and
hear the roots crack as they pass below, delving along very
slowly, though they can *run* very swiftly in their galleries—
so swiftly that you can hardly *see* them, which is not the
case with them outside, for dry weather drives the young
forth for food; and the Jacks, too, come forth to fight, and
fight fiercely with their claws, with their fur laid back, and
their bead-eyes staring, looking twice their natural size.
And you may know they are males, for they are larger than
the females, have bigger feet, and many of them have a
yellowish spot under their bellies, probably an age-spot,
or "grey hairs." These must not be confounded with those
cream-coloured moles—found in certain soils by Flegg and
Reedham—albinos not uncommon. And should you catch
a mole just as he comes from his run, you will find him
covered with fleas; and, strange to say, directly he gets
into the open air, the fleas leave him, just as vermin-like
ideas leave men when they too get into the open country.
They will leave their cover too sometimes during snow-
storms, and draw to thinned patches, "rutting" like a pig
in the grass, occasionally skimming a few inches of turf.
They can swim, too, like a water-vole.

They do much damage. One will skim down the middle of a mangold balk on a hot day, and "share them out." as cleanly as a plough would, the roots dying and withering. They play the mischief, too, in barley-fields, for they prefer soft ground; it is easier work; but they will work anywhere. On the uplands, however, they do not throw up hills, but merely place their nests in a hole in a hedgerow, each family having a hole to itself.

When in the fields, they will soon leave if there is anything they do not like—a poisoned companion, for example—and wander off to some neighbouring field; indeed, they will "work up" a field in a few days, so old mole-experts assure me.

In the spring-time the keen trapper is busy setting his primitive snares, rubbing the bows carefully with mould first to kill all scent, and then being careful to set the trap in a main-run; for it is no use, old trappers say, to set the trap where they feed, but only in the main-run to or from their feeding-galleries. They also frequent the main-runs to river-wall or hedgerow when the ground is very damp. I have seen one trap catch after it had been set five minutes, but such is rarely the case. A catcher gets eight or ten from one main-run, or thirty or forty if lucky.

When caught, they are always gibbeted on thorn-sticks, to show the farmer's eyes and please his heart. I have sometimes seen a hundred and sixty carcases hung on a thorn-bush, in various stages of decomposition.

Some mole-catchers say they put a drop of oil of aniseed on the trap to attract the moles. But one old fellow told me "he once tried that, and the warmin sarved him out." He said he put a drop on the bows, and caught one; then he put another drop on, but when he went to the trap again it "weren't sprung;" and on examining it, he found they had "blocked" it, *i.e.*, filled it with earth. This old fellow used to put young moles on a bench, but they could see it had a fall to the ground, so they never went over the edge. They used to bite him sometimes, but never to hurt.

Latterly, trapping is being supplanted by poisoning. Thread is soaked in strychnine, and worms are threaded and put into the runs. Experts tell me it is not successful in killing the moles, though it often drives them elsewhere.

Besides man, the mole has many enemies, the most formidable of which, especially to the young, are crows, rats, weasels, and stoats, for they will all eat moles, and all three take possession of the borings afterwards as well, young rats often being caught by terriers in mole-runs.

And lastly, the unscrupulous mole-catcher, who owes a farmer a grudge, will quietly go to a nest in season and take some well-grown youngsters into his pocket, and turn them off on the farmer's land, where the young hopefuls begin "skimming."

And such is the story of this wonderful little animal, that builds his home *with hands*—and not without them, as the garrulous and inconsequent have stated.

CHAPTER V

THE OTTER

THE largest and fiercest wild animal of the Broadland is the otter, an old dog measuring four feet, and weighing from twenty-eight to thirty-four pounds, and to this weight are added the jaw and courage of a bull-dog and the teeth of a tiger.

His habits are simple. He builds himself a nest about the size of a swan's nest (occasionally in an old swan's nest), in a hollowed-out hole, in thickets of sallow and bramble, in thorn-bushes or old furra (furze) bushes near a dike, in thick rushes; but floating reed-hovers near pulk-holes are his favourite nesting-places, for there is a handy trap-door into the water beneath, a ready means of escape—for on hearing the slightest sound (and nothing is quicker of hearing than an otter), he dives down, swimming off under the hovers in safety. Sometimes you may find a hole in the bottom of his nest through which he comes and goes. The nest is open, no roof being built over it, though he often pulls the reeds and stuff over in a careless manner. The nest is lined with litter and reed-leaves, and is warm and cosy, as any one can attest who has placed his hand in one on a cold winter's day just as the otter has left, for they lie in or by their nests, or on the hovers all day, often sleeping; but, as a rule, they are light sleepers, though I know one fenman who caught one asleep, and had time to go and get his quant, and return and "leave his mark on him;" but the otter finally escaped.

Opinion seems divided amongst the Broadsmen as to the date of younging. One expert assures me they young in

January, and he has seen two, three, or four young pups on the ice in that month; others aver they breed in May and June. I have never yet seen the young *in* the nests, but have seen young ones caught in May, but could not tell their age; so I must leave this point open.

In winter the coat is darker and longer, and covers the shorter, lighter, buff-coloured hair. Still, on warm fore-noons in February, I have seen them on hovers sunning themselves, their winter coat all shot, making them look as white as snow in certain lights. But they are more commonly to be seen at closing-in time than any other hour of the day, just as they are starting off to fish, or "work," as the fenmen say. And they go miles on these expedi-tions, across bog, water, and land. I know one trail, about nine inches wide, that leads from their lairs along the ronds (they do not frequent dry marshes) bordering a river for a couple of miles, the trail sometimes running into the shallow water (for they prefer a shallow broad for fishing, and you may hear them on a high tide breaking like tigers through the reed-beds) and out again, on across dikes, a pile of green swan-like dung, with fish scales, marking the place he got ashore (for they generally dung after a swim, the water they swim in seldom being more than four feet deep), on across wet marshes, over dikes, and to a shallow broad, where he fishes the greater part of the night, returning at daybreak. I say he—but a colony lie in the reed-hovers I speak of. And if you follow their trail early of a morning, you will find remains of their meals, great eels—some weighing as much as four pounds—half-eaten, the bite being as clean as if done with a sharp knife; but if we may judge by his leavings, the otter likes eels least of all, and, by the same standard, he likes tench best of all, eating the whole fish right up to the gills. But with other fish the case is different. I have seen eels with only a third of their bodies eaten, others half-eaten, generally up to the middle; but I have never yet seen the classic spot " behind the shoulder" touched. Rudd, roach, and pike, too, are eaten, and bream. I remember a fenman once telling me excitedly he had seen an old otter with a great old bream in his mouth, and he informed me it took the fish under water. I have seen a pike eaten in the classic spot—the shoulder.

In winter their "feetings" or foot-prints—about the size of a man's fist, with closed fingers—are to be seen on the snowy marshlands, and in their trail the wings and claws of water-hens, coots, and wild-fowl—for they will eat all these, taking the wings and feet off as does a cat—also fresh-water mussels; and I know one fenman who saw an otter eat a nest of young wild ducks. The otter's mode of progression is interesting. His legs are as short as they are stout, and he seems to glide like an eel over the wet marshes or ronds with dragging tail; but when opportunity offers, they will shut their fur, shake themselves, and roll over and over like a dog. When frightened, their tail spreads out.

The eel-catcher and flight-shooter see more of him, perhaps, than most of the Broadsmen, for they are on the alert at closing-in time and at daybreak, when the otter draws home. Should the otter not get home by daybreak, it will lie up in a deserted shed or mill, or roll itself up in the thick stuff till night. One eel-catcher tells me he was setting his eel-net in a dike, when he saw something move close to the shore. He stood silently watching the spot, and suddenly it flashed on him there was an otter. So he dropped his net and went into his boat-house, and got his "old Betsy," and crept along the shore. The otter saw him, and slipped into the dike; and as the sky was clear to the westward, he could see the otter when it put its head up to breathe. When the otter came up a second time the eel-catcher fired, and under he went. Up it came again, and again the eel-catcher fired. Under it dived again. They kept up this hide-and-seek game for ten shots, the otter going about forty yards between each shot. At last the otter got tired of playing the target, so it drew out on to the shore. The eel-catcher at once dropped upon his knees and fired, putting the whole charge into the animal's back. Going back, he got into his light marsh boat, and quanted down to his prey, finding him dead. Some of the fenmen say "they take the flash off the gun." A few nights afterwards this same man heard one in the reeds, so he set a large steel-fall in the brute's trail, placing the trap under water. The next morning he went and found the otter had pulled the trap into the reeds, where he discovered him. He struck at the brute with the stock of his gun. The otter dodged the blow, and dived under the hover. As he came up again, the eel-catcher struck at him again; but again he dived and escaped, and it was not till the fifth stroke that he hit him a smart blow, the otter seeming apparently dead. But as he went to pick him up, he saw the creature move; the beast was shamming. So he struck him again, and again the brute shammed, until the third blow finished him.

Once when flighting I heard a soft whistling in the reed, a soft *wheet, wheet.* I took the noise for a spotted rail; but the season was wrong for them, and as I peered into the water, an otter passed within two yards of the boat, a-fishing bent, no doubt. Every time his nose came up he whistled. A gunner told me he was flighting one night, when an otter passed within fourteen yards of his boat. He waited till the beast swam into "a clear," when he could see the creature fully—his head, part of his body, and his tail being above water he fired just as it went into the gladen, and on going to the spot found strips of felt two inches long, but no otter. This same gunner has shot two out of five whilst playing in the gladen, and says he once saw a big "bull-headed" one—"the biggest warmin he ever see." But this must have been due to some peculiar effect of lighting, for he assures me "the warmin's head warn't flat." Otters look darker, too, just as they leave the water.

Perhaps one of their most astonishing habits is their method of progressing along a frozen dike. They swim along under the ice for fifty or sixty yards, and bump their heads up against the glaze, breaking a hole from seven to ten inches in width, and this in ice up to three-quarters of an inch in thickness. Should the ice be clear and brittle, these blow-holes will be larger. In wintry weather, it is evident by their trails that they seldom go more than twice or thrice in the same direction, for they soon tread a path—half-a-dozen journeys being ample to lay a path.

And the most poetical time of the otter's life is mayhap in the early spring, when he lies in his warm reed-leaf nest, cosily hidden in the hot moist thicket of budding sallows and white-sprayed blackthorn, listening to the rustling of the tall, pale, dead grasses round him, or to the buzzing of the black and yellow bees, and the blue-tits' song as they hunt over the sweet-scented sallow buds for pollen and insects; and mayhap too they hear the red-legs calling from afar across the level green marshlands.

CHAPTER VI

POLECAT.

I HAVE never seen a polecat in this district, except gibbeted in some cover. Nor are they common. But they are occasionally trapped and shot. The last I heard of was caught in a rabbit-trap, and I was told "he was bigger than the biggest buck ferret, with a sharp nose. He was savage, bit at the traps, and was the colour of the polecat ferret." I heard of another being "muddled up" beneath a low bridge over a water-dike and killed.

THE STOAT,

Or "minifa" (minerva), or lobster, on the other hand, is common in the marshlands and in the covers. 'Tis just the country for him, for he loves the water almost as well as the land.

You may see his brown and yellow fur gleaming on the rush marshes in early spring; and should you start and run after him, you will see him bound across the land, and suddenly disappear into the ground, and when you come up, you will find he has taken to a mole-run. If you be patient, and job about with your stick in the old mole-heaps near, you may find the fur of a marsh-mouse in a dry and cosy old mole's nest, for no sanctuary is dearer to him. He turns the moles out of their home, and makes his quarers there, feeding upon marsh-mice, eggs, young birds,

344

and chickens. Should you trap him in a mole-run (as is often done), you will find he measures eighteen inches in length, and is about the size of an ordinary polecat ferret, the female being a little smaller. In winter, of course, he is often white, with a black tip to his tail—hence his name of "minifa." Whether they pair or not I cannot tell, but I am inclined to think not. The nest is often made, as I have said, in a captured mole-heap, as also in a *dry* hedge-row or river-wall, or old mossy bank, or in a hollow tree; and the animal, like the weasel, betrays its presence by leaving a pile of dung at its front door—a clean practice, but an unwise one. They will nest, too, in a heap of old rushes, that has been "laying" for some time. In April they young in their nests, composed of mixed feathers, mouse-fur, and grass, and the old animals find plenty of mice, rats, rabbits, and leverets at this season to feed upon, not to mention eggs and young birds.

Their prey is hunted purely by scent. Sometimes they hunt in couples, at other times in packs. One old fenman tells me he once met a troop of the "warmin" going along like a pack of hounds. Once I came across one hunting a rabbit in a planting by the water-side. The rabbit came madly across a plank laid across a nine-foot dike, and the stoat, with ears pricked, with nose set out, tail rigid as a bar, and coat all standing, on his trail. About half across the plank the stoat made a leap and alighted on the rabbit's back, the frightened creature rushing across with his dread foe on his back; but the keeper's dog frightened off the stoat and secured the dying rabbit, for it had received its fatal wound. But the stoat left his disgusting smell in revenge, as they always do when "stowed up." His fights with rats I shall elsewhere describe. He is just as great an egg-stealer as the rat too, rolling them along in the same way to his hole in the river-wall. When he starts on a hunt, he begins like a hound, but with nose closer to the ground, snuffing the trail; and should he lose it, he will fly round and snuff about, coming upon it very quickly. When he fairly strikes the trail, he goes along at a trot, often stopping and looking all round him to see that the coast is clear. In this position he resembles a snake. As he comes up to a rabbit it seems to get fear-paralysis, and limps along or stops dead, screaming a moment after as the sharp teeth seize it behind the ear. A fenman once caught one in the act of sucking a leveret, and he caught the pair—a youngish stoat and leveret—and put a string round their hind-quarters and carried both home. When he separated them, the stoat sprang at him, and hissed and made water, which smelt abominably. When the stoat has sucked the victim's blood, he will carry it away (if not too large) to his hole; if too large to drag, he will leave it and cover it if he can, but if he is not hungry, he leaves it like a dog, and many a marshman has dined off rabbits and leverets killed by "them old lobsters."

At times, if hard pressed, the stoat will attack a full-grown hare, flying on to its back when it gets well within reach; and marshmen tell me they have seen the poor frightened hare running through furze and under gates to try to shake off the terrible foe, who was gripping his throat tightly with all fore-feet—feet more rat-like in shape than an otter's. Others have seen stoats go round their prey in ever-narrowing circles, the poor rabbit sitting unable to move—both of them shrieking until, with a final spring, the tragedy was finished.

That they do kill game is certain, for I have seen them take a young partridge, that could just fly, from a clutch. On being frightened the stoat skulked away, but soon came back, and began eating at the head—like birds of prey, they prefer the eyes and brain to any other part.

They only have one litter a year, like ferrets; but a ferret has nine or ten to the litter, whereas a stoat has but five or six. When the naked and blind young things are born, the mother suckles them until they get their sight and can hunt for themselves, which they begin to do at a very young age.

Still she keeps suckling them after they can eat for themselves, until they are quite able to take care of themselves, when she weans them, and discharges them to look out for their own food, when they are almost as big as their mothers. An old rat-catcher and breeder of ferrets assures me they will "lay and suck meat before they can see," just as ferrets will. He says he has split a sparrow and laid it down by their nests, and they have drawn out before they could see—working by scent—and sucked it.

Should you disturb the nest, however, when they are very young, the mother will kill her litter; and should they stroll forth of their own accord to see the world, the mother will catch them by the back of their necks and carry them back.

Like ferrets, too, they will not work much in damp weather, for the scent is bad. At the breeding season they will go into the water after rails' and water-hens' eggs, and young birds—for they swim well, dog-fashion. They suck the eggs by breaking a hole in the sides with their teeth.

The males fight one another fiercely at the breeding season, setting their backs up, shooting their tails straight out, their bodies swelling with rage, and their fur standing on end; flying at each other's heads, gripping each other, and rolling over like dogs, then separating, running after each other, jumping over each other, and hissing; now gripping each other again, and shrieking and cackling like a ferret, making a frightful stench all the time, until one is worsted; for they seldom kill each other in these duels, though that does happen at times.

A favourite haunt of theirs in summer is the warren—in an old rabbit-hole, whence the marram-cutters often disturb them in July, as they mow the sandy crop for thatch and litter. It has been said they frequent corn-stacks. This is not my experience; and I think the weasel is the corn-stack hunter.

The rat-catchers find stoats difficult to trap; but they do catch them, preferring to set their traps in a drain-pipe under a gateway—a favourite haunt of the stoat.

THE WEASEL, OR MOUSE-HUNTER,

Differs from a stoat as a mouse does from a rat. He is smaller, quicker, less fierce; otherwise, his habits and customs are much the same as those of the stoat. But he has no black tips to his tail, and is only seven inches long.

He frequents corn-stacks, and old mole-holes in the river-walls and hedgerows, as well as deserted rabbit-holes on the warrens; for he loves dryness and warmth, and is not such a lover of hunting by water as the stoat, though he can swim as well. When running—and I have seen them scamper for yards along a river-wall, for they never leave their nests for long—he resembles a stoat; but when he stops, he does not stretch out his neck like a snake, but seems to turn it between his shoulders more. Indeed, altogether, in build and action he is more rat-like than the stoat.

The nest, similar to a stoat's nest, is built in a hole in a warm river-wall (dung betraying his presence), or in a rat's-run, or at the bottom of a stack, and five young ones are born there, and live and grow as do young stoats. But the weasel nests earlier than the stoat.

The weasel, too, hunts like the stoat, catching and feeding on mice, young birds (including game-birds), if they get separated from their parents or are weaklings, very young rabbits and leverets, and rarely a rat; but they will, on occasion, fight a young rat, and win too. It is not probable that they will take birds from old pheasants or partridges, and may therefore be considered innocuous to game, though they are more cunning than the stoat, being very swift, and hardy too. Mice and voles are their chief food—mice by the farmsheds, and field-voles on the marshes. Also, they will suck *small birds'* eggs.

In stacks they are all alert, and will draw under the stuff and thrust their heads suddenly through the thatch in the spot they think the **bird** is sitting; for they, **too,** work by scent, like the **stoat.** They will hunt little rabbits when they get separated from the does, swimming (and **they** swim much **faster than a rat) up** the **dikes to** circumvent **them** if necessary, or **hiding in the grass until they pass closely to them, when they will pounce upon them.** An old fenman once saw a **good-sized** rabbit carry a **mouse-hunter on its back for over three** hundred yards.

Should you catch him with his prey, generally a mouse, he will immediately drop it; and if you "hide up," and be patient, you will see him return and pick it up again, taking it to his burrow, even if he has to carry it a hundred yards.

No place seems inaccessible to their long, slim bodies, and they can enter almost any crevice or hole that will allow a mouse to pass. And should you "stow them up," they will emit the same stench as a stoat, though it is not so pungent.

In the spring, too, they may be seen fighting, after the manner of a stoat, on the **sere grasses.** Old rabbiters tell me if **a mouse-hunter** gets into a **sandpit** full of rabbits' "**eyes,**" **he will stop** there all summer if not killed, whereas a stoat will not. These men say, **too,** the **mouse-hunter** "**kills for the love of** killing." However, he is a **harmless** creature, and **should not** be killed by the gamekeeper.

Gilbert White, and many of his imitators, talk of a smaller species of weasel. I do not believe there exists such a species, but that the female weasel, which is smaller than the male, is responsible for this mistake; and the looseness and dishonesty of his plagiarists have perpetuated the error—most recently confirmed (?) by that inaccurate "naturalist labourer"—"Son of the Marshes."

CHAPTER VII

THE RABBIT

In the Broadland there are two kinds of rabbits—the hedge-row rabbit and the warren rabbit.

The hedge-rabbit is a destructive creature, and a most prolific breeder, breeding in mild winters every month of the year (February, however, being their chief month), and the doe taking the buck the same day the young are cast. They make special holes in the banks and open fields to breed in, their holes being often two or three hundred yards from their homes. Before casting her young, the doe will almost strip herself of down, with which she makes her nest, mixing it with grass. In this warm, downy nest the young, numbering from six (the average number) to ten, are cast, their birth being generally by night, the blind, feltless little creatures being left by the mother in the day-time, when she covers them up with down, for warmth and security. They remain blind for nine days—if they live so long, for sometimes she deserts them, and is said to kill them if any one disturb the nest. When they are three weeks of age she will leave a small entrance to the young, and when they are a month old they come forth into the great world and fend for themselves.

And then she is often already a fortnight gone for a fresh litter, and on the look-out for a fresh hole, if her first nest be in the open field; if, on the other hand, it be in the hedgerow, she may throw them off to do for themselves at three weeks of age. Her felt grows rapidly, though it keeps thin in summer-time, and if you hold her up to the light you will see her bare skin. She is more free from milk in the summer too.

Apart from the breeding, hedge-rabbits live in their burrows, often placed in a hedgerow. They always have one gallery to escape by, but some have many more, certain galleries being blind alleys. The number of "eyes," too, is no indication of the number of rabbits in a burrow; for after searching twenty or thirty holes with ferrets, perhaps only a brace of rabbits will be found. Their galleries are about three or four feet long, and turn in all directions—the rabbit sometimes being protected from an enemy by only a few inches of crust, for his burrow may work upwards, being really but an inch deep. When the hedges are thin they take to the open fields for their burrows, being indifferent to the soil in which they burrow, be it sand or clay.

The young soon begin to nest for themselves. One old rabbiter told me he once caught a young rabbit about three-quarters grown with four young ones inside of her. As a rule, the first two litters of the year breed the same summer they are born—the first litter probably breeding twice.

During the day they either "lay up" in their burrows or steal forth to feed, never going far, however, from their "eyes" in search of food, thus differing from the hare. But they never approach their young by day, but steal off to them at dusk.

The old bucks will fight in the same way as do hares, knocking each other over, and wool-gathering, and you may see them gambolling, copulating, and otherwise enjoying themselves; and should you disturb them, you will hear a sharp signal of danger given by some old buck—a sharp knock on the earth with his hind-legs, and away they all scatter, for a rabbit can run as fast as a hare *for a short distance;* and all you see as you come up is a patch of white disappearing down a dark "eye." At night, too, those who are not nursing go forth to feed, never straying far from their burrows even then—two fields' distance at most— returning, however, soon after dawn to their "eyes." If the night turn wet or cold, they return earlier; and if it be cold or wet at evening, they generally lie up all night, and do not go out at all.

As the corn, and roots, and beans grow sufficiently high to afford cover, they leave their burrows and live in the fields; especially fond are they of bean-fields. In these fields they make forms after the manner of a hare, and they will elude a dog, if they know where their burrows are, for a distance of one or two hundred yards. It will take a very swift dog to catch them, unless they get confused or baffled, which sometimes occurs. I have heard that for a hundred yards they have outrun greyhounds, but I never saw such a race.

Their food much resembles that of the hare — young clover, young wheat and barley, sow-thistles, red-weed (poppies), turnips, beet, cabbage, and young greens and salads, and in hard weather hay.

But a rabbit shows the extent of his damage, for they do not feed far from their burrows, and you can see what they have eaten and what they have killed by their poisonous excreta.

Warren rabbits are larger and more sociable than hedge-rabbits. They seem to congregate more, and breed in colonies. Their burrows go deeper into the sand, some nine or ten yards, and they go straighter in. They eat marram and rushes. Often, when they are killed, you find sand in their lips, and it seems as though they use their mouths as well as their heads in burrowing. Indeed, you may see them washing their faces like cats, but you may also see hares doing the same. They are a hardy crew those "warrant rabbits," but a family of which I have but little experience.

In habits, however, they much resemble the hedgerow rabbit, except in the points noted.

And finally, a good young rabbit makes one of the best dishes to be had, an you cook him properly.

"ON GUARD."

CHAPTER VIII

RATS

THE most cruel, most destructive, and most hateful of all the vermin of the Marshlands are the rats.

Experienced rat-catchers of the Broadlands distinguish three kinds.

1. The largest is rabbit-coloured, with a yellow chest.
2. The next is the large brown rat, verging to a red.
3. The smallest is the little red rat.

1. BIG RAT, WITH YELLOW CHEST.

Experienced marsh-ratters assure me they have killed tnis formidable creature two feet in length, and weighing two pounds. I myself saw one killed that measured twenty-one inches.

This fierce, brave, and detestable animal frequents barns, corn-stacks, and hedgerows close to buildings, often resorting to the warrens in the summer-time.

In March, when the reeds rustle and the marshland is an arrangement in blue and gold, the old pairs begin to build their nests of the handiest materials—grass, straw, wool, paper, and rags all serving their turn. The nest, "like a rabbit's nest," is usually placed in a roundish hole near some warm spot, such as a bullock-shed or heated manure-heap. And the wise old Jack and his partner are careful that they have several exits and entrances, four being the usual number, though I have seen a nest with only one entrance; but that was incautious. In this cosy nest the litter of fifteen ratlings is born, and very fierce are the parents at this season, fighting unto the death for their bantlings, though the Jack is not commonly to be seen after their birth, the mother suckling them, and defending them, for the Jack rarely lies in the nest with the females. Later on, as it grows warmer, the nests (for they young every six weeks from the beginning of March to the **end of August, and at intervals throughout the year, young rats being occasionally found in frosty weather, but such is unusual) are placed in cooler places, under barns or ricks. When the young are weaned, the old Jack becomes more assiduous, and both the parents feed the youngsters, being** very **fond of a suc-**culent young chicken or **duckling for the** purpose, sometimes robbing the millman, fenman, or farmer of fifteen young ducklings of a night, dragging them into faggot-stacks, where they will be handy.

And when the young litter is full grown, the old parents drive them away to fend for themselves, which accounts, perhaps, for some of those curious migrations one comes upon in the marshes occasionally—a party of rats on the march, the largest and fiercest leading the way to some un-rat-frequented barn or hedge. One **old fenman once met** between thirty **and forty** of those "**warmin on their trams, and they went along as unconsarned as passengers, and wouldn't budge for him—not they.**"

And a hungry horde they are, eating anything—grass, **turnips (the chips** betraying their presence), eggs (which **they roll** before them as does a stoat), young birds, dead **fish, frogs** (which they disembowel, leaving the gaping torsos in the ronds), small birds—all is game that comes to their teeth. I remember one hard winter seeing one of their nests filled with larks' wings, the birds having been caught, no doubt, whilst roosting on their grassy forms on some moonlit frosty night. A long way will these devastating hordes wander in search of new quarters, across dikes and marshes, up walls, and over roofs (nothing stops them), perhaps to **take up their residence in** a rabbit-burrow, where they live with the rabbits in the same family, finding plenty of tasteful food at hand.

And won't they fight, these old brutes. I know one rat jumped at a ratter who had "cornered" him, and ripped his shoe open, nor would he let go till he was killed, still clinging to the man's foot. But men, dogs, and ferrets are their deadliest foes; for the man, if he do not shoot him, puts his ferrets in, though he is cautious how he intrudes his ferrets into the holes of these large yellow-chested rats, for often the ferret is killed. The rat lifts itself up, pressing against the top of the hole, and jumps down on the ferret, cutting its throat sometimes as neatly as if done with a knife. Indeed, in a stack they will sometimes turn the ferrets out. One old ratter tells me he once knew some of these same big rats to be under a granary floor. There were three feet between the floor and the earth, and these rats had stolen seven or eight coomb of corn, carrying it through a hole in the floor, for they are very fond of corn, and can be seen inside of granaries, sitting up against the windows licking the dew off, for they cannot live without water. This ratter was sent to kill the thieves. He pulled part of the flooring up, and stopped up all ways of escape, and put in fine big, strong Jack ferrets, and waited two or three minutes, the silence being broken by a regular stampede of these creatures, who ran into a heap in the corner—the pile of living rats being, he says, "nigh t'ree foot high." One dog flew in and bit a few, killing none, and they scampered back and fought the ferrets, killing one outright. In the meantime the ratter got his gun and his elder dog, and as the vermin came forth again they killed between them twenty-seven monsters, though the brutes drew themselves up on their hind-legs and flew at the dogs.

This old man could catch a rat neatly. He used to nip the tail between his left thumb and finger, and pounce down on the nape of the rat's neck with his right finger and thumb, and he had the beast securely, as one catches a ferret.

One of the best matched duels is that between a stoat and a rat. Let us watch such an one. The rat bounds clumsily away across the clods, his tail straight out, and his jolty run availing him nothing, for the stoat quickly gains upon him, when the rat stops and jumps on his hind-legs, his head raised, his ears a bit back, and showing a little ivory, but he never seems certain of success —the stoat, with erect tail, and ears and hair shot out, approaches him slowly, turning his head now this way, now that, and suddenly darts upon him, seizing him by the back of the neck. Sometimes the rat squeaks and is dead at once, and the stoat shakes him, puts him down, but never drops him, and holds him between his teeth, apparently sucking his blood for some minutes, when he will suddenly look round and gallop off to the hedge with his prey, to hide it. At other times the battle is longer, and the combatants go rolling over and over, scuffling and squealing even for a quarter of an hour, until, with a shriek, the rat gives up, for the stoat always masters him in the open. Sometimes the rat is driven to the water (in a dike), for he will never take to it unless forced, for the stoat is his master there too. Should he, however, be forced to a dike, he will boldly plunge in and swim under water some twenty or thirty yards, coming up cautiously with his nose just above the water; but the stoat is after him with his otter-like strokes, and the rat swims off again, for he will never fight in the water, and the stoat is soon up with him, and all is over. Should a rat be "stowed up" in a pit, and he sees no chance of escape, he will sooner let himself drown than come ashore to be killed. And well he may; for some of these mongrel ratting dogs are wonderful. I know of one who will kill a rat one minute and retrieve a delicate bird the next, without disturbing a feather; so clever are some of these ratters at training their dogs, and so clever are the beasts. In summer, of course, little ratting is done, and the ordinary ratting practices are too well known to repeat. But on the Marshlands, a moonlit winter's night is a favourite season for this sport.

When the marshes are white fields and the dikes and rivers are leaden, and the moon shines down on the white powdery landscape, the ratter and his mate, well muffled up, will go with their dogs to the farmer's ricks, one going round the rick one way, the other taking the opposite path, both the rat-catchers kicking the brittle straw sharply, their warm breath coming and going in the cold, clear night, as their dogs stand all alert, when suddenly a rat jumps from the rick through the moonlight to the snowy carpet; the dog yelps, there is a scuffle in the snow for a moment, a shriek, and a fierce shaking, and the rat drops dead and crumpled from the dog's teeth. And in two hours' ratting by moonlight experts can kill as many as they do in a day. Finally, it puzzles the adept ratter to know why he never kills a young rat by night; but such, they aver, is the case.

2. Large Brown Rat.

The rat already described may be called *the* rat of the Broad farms and outbuildings, but the large brown rat is *par excellence* the rat of the marshes, broads, and rivers, and dikes: these are his hunting-grounds—the fields and marshes; therefore is he the farmer's most dread enemy.

For his headquarters are in the "walls" and hedgerows, his foraging grounds the mangold, turnip, and corn fields. He is especially destructive to wheat, barley, and turnip crops whilst in the ground, for he never goes near the barns; those are the hunting-grounds of his brothers already described.

In March he makes his nest in a mangold balk or "holl," and the female is pregnant about a month before the dozen youngsters are born. These, too, breed every six weeks, and go on breeding till the end of September, more rarely having a litter in winter than his barn relatives.

The nest is a large, neat, round cavity, made of straw or reed, and the litter stuff used for covering the "beet." Later in the season, when the weather is warmer, these rats prefer to nest in the hedgerows, or dry "deeks," having one or two escape-holes. Here they will rear their young, fighting any interloper to the death, for this rat too can kill a ferret on occasion. It likes the water better than the barn rat, and will in hot weather lie by the dike-side basking. They feed near the water's edge, too, in early spring, when other food is scarce, diving to the broad bottom and getting the "clams" or fresh-water mussels, of which they are fond, taking them to the shore to eat. Their trail, a few inches wide, can be followed for half a mile along a broad, or they will go straight across the marshes to some friendly stack. If in March you wander along a broad, you may find such a trail about a foot from the water, winding through grass jungles, sedge tus- socks, over rotting alder branches lying on the broad beach like mangrove stems, for the alder is the English mangrove. On goes the trail across quaking bogs, where the soft earth yields to your footsteps, and quivers as you pass across. Every now and then on the path you will come across a midden heap, a pile of dry and polished mussel shells, and gnawed gladen roots, lying amongst rotten alder branchlets peeling in the dry air. Mayhap an you follow far enough, you will find the run turns to the water and leads over a polished alder root into the broad—no crumbling platform to take off and climb upon, but a natural landing-stage.

In winter, should you lie in your boat by the reeds, you will soon be visited by these vicious creatures; you will see them running up and down your mooring-ropes, and hear them scampering across your plankways, when you lie in your berth trying to sleep. And if, in disgust, you go forth in the moonlight and shoot them with your rifle, you will see them, if you watch, carrying off their dead to eat, for they are cannibals; or mayhap you hear ducklings cry in the dike as they take them, and you may perchance find a duckling even to two months of age lying by the dike-side, with a hole large enough to admit your little finger bitten in its throat. And if you wish for revenge, you set your steel-fall baited with a duckling, putting the trap under water, and it may be the next morning you will find your thief looking mournfully up and down the dike.

I remember an old keeper who loathed rats, and his hatred began in this way. His cottage was in the heart of the swamps, in the midst of the jungle. One day he wanted to get six brace of snipe for his master's table. Going forth, he got two brace before breakfast, and laid them on a bench before his "clear," whilst he went indoors for his meal. When he came out, they were gone. He found a wing in the rushes, and "laid that to an old rat." Taking his gun, he went forth and got seven and a half brace of snipe before the night. On his return to his cottage, he dressed the birds, and laid them on the table in his room—it was a one-roomed cottage. About one o'clock in the morning he was awakened by a noise at the door; he got up and went to the door, but saw nothing. Coming back to bed, he turned in again, and was again awakened about an hour afterwards by the same noise. Rising, he went to the door, and as he opened it a great rat flew past him into the room. Shutting the door quickly, he lit his candle, and arming himself with his poker, made for the rat; but the rat was too sharp for the old fellow, and kept dodging him till he was shining with perspiration, and well-nigh spent. At last the rat, being pressed, jumped into an old tea-chest, filled with nets and liggers. As the old keeper moved the chest, the rat jumped out, and got into the corner, imprisoned by the chest, where the poker finished him. But his daring and wolfish nature is well shown by the story—he wanted the snipe.

In the prime of summer-time, when the sandpipers call in the dikes, these rats will go down from their burrows, and lie by the water, "laying up" in the cool dampness of the dike. They travel in droves, too, when full grown; and an old fenman told me he met from thirty to forty of them going across the marshes in Indian file, a fine big fellow leading the way. Nor do they confine themselves to travelling the marshes; but, when half-grown, will run up the rat-catcher's limbs for shelter when hard pressed. The many disembowelled frogs to be found on the ronds and marsh-walls in early spring are the victims of these fierce marauders. And should a ferret find a frog in their holes (as he sometimes does), he will eat it as well as the rat.

Their deadliest enemy is the stoat, who attacks them in the same way as he does the larger barn rats, and always masters them, the rat's open mouth, thrown back head, and arched back being of no avail.

When the corn is up, they are to be found harvesting with the men; and should there come a drought, followed by a soaking rain, you may see them turn out to drink. When the hard frosts come, they will lie up in their burrows, seven or eight lying together, or else they will seek the warmer rabbit-holes on the warrens or in the sandhills.

A novel way of trapping them is to place a "red" herring's head on a wall a foot above the trap, when they are sure to get caught as they "reach for it."

3. LITTLE RED RAT.

These rats are much rarer than the two larger kinds, and go by the name of "Italian rats," and "ship rats;" for old fenmen say they come from foreign ships wrecked on the coast. They are generally found in "schools." This rat is about a foot long, and frequents banks. Old rat-catchers tell me he doesn't "young" till late (never earlier than May, until which season they often remain in bunches laid up in holes), and that they are great cannibals, eating each other's young. Indeed, one rat-catcher tells me all rats do this, more or less; but the red rat is fondest of his neighbour's children. They are a pest to the farmer, too, robbing the barley and wheat recently set, following the drills, and completely ruining the crop, pulling the corn into the hedgerows for better consumption.

Their nests—made of old dead leaves—are said to be found in the hedgerows; and when they get a good size, they are said to take to the corn-fields. I am told they are fond of building their nests in the laid corn, or " by water-dikes in thickish stuff." But the stoat is on their track whenever he can be, for he seems to love to kill a rat beyond everything. And, like most rats, I am told they get sores, and grow mangy towards the beginning of summer.

Rats of all kinds often get into a wherry, and they will stay there if they can work from one end to the other, for they never eat waterwards. I know an old wherryman who once caught eighteen rats from a wherry by scattering pollard, with *gloved hands*, over a trap, and on a plank leading to it. The first night the rats went a little way up the planking; on the second, a little way farther up; and the third night they essayed the pan, and the first got caught in the steel-fall.

It is a pity this cruel, wolfish, destructive family cannot be blotted from the land; for they work nothing but evil, and bring misery and destruction to many a happy animal and bird—they are the criminals of beast life.

CHAPTER IX

SHREW MOUSE, OR RANA, AND SQUIRRELS

THIS strange little creature is not uncommon in the Broadland ; and you may often find its nest in the marsh bottoms, with four or five young shrew mice inside.

The shrew mouse is a frog-killer, and has a peculiar love for cinders, which he will carry away from the cottagers' fireplaces. He is useless for eating—even a cat will not touch him.

SQUIRRELS

Are not uncommon in the plantings on the broad edges, where they may be seen feeding on the deal-apples, as the Broadsmen call the fir-cones ; nuts and acorns too they are fond of. On snowy nights they descend from their trees, and you may see their "feetings" the next morning all about the carr. Should you catch one, you will find him full of fleas, especially between the legs. Stoats sometimes chase them, but "pug" generally manages to escape, an he can get to a tree first—he is a much more daring jumper.

CHAPTER X

BREAM

THE "great old bream," as this fish is familiarly called in
the Broads, is a bottom-feeder—a real, rootling, pig-like
bottom-feeder—eating the " soil " at the bottom of deep holes ;
for he too likes deep water, especially in winter, for it is
warmer there.   Should you examine his stomach, you will
find a black, tarry-like substance—weed and mud.

Bream, too, keep in shoals—the large fish keeping in shoals
to themselves, and the small fish doing the same. The bream
collect for rudding in similar places to the roach, but, as a
rule, a few days later, though at times both shoals spawn
together, and all the remarks on roach spawning apply to
bream spawning.   Some old fishermen of the district main-
tain stoutly that the rudd is a cross-breed between the roach
and bream, born from their commingled spawnings ; others
aver as stoutly that the "bream flat " and " white bream "
are nothing but young bream, and try and convince you
by assuring you they have never found "full" white bream,
" but that I must leave."

Upon a still, warm day you may peer into the liquid
depths and see the old bream working on the bottom,
wriggling and rootling in the mud, clouding the water
with silt till you cannot see them, though they keep on
feeding, working from side to side, their noses under .the
mud, as you may tell by the ever-thickening water, as they
hollow out the bottom. Some say they are hunting for
worms ; but I have never seen aught but the black, tarry,
muddy sort of soil aforesaid in their stomachs.

They, too, feed mostly of a night, though they feed by

day, but there is no fixed eating time for a bream; but once
he starts feeding, he is likely to go on, if the shoal be kept
together.   Should a bream catch sight of you, if in a shallow, he will dart off, and "mud," reappearing later on.

A fine old four-pound black-backed bream makes a grand
dish if you catch him when full, some two or three weeks
before spawning, and cook him *à la juive*, as described for
pike and roach.

Old fishermen tell you, however, the way to cook a bream
is to skin him, and let him simmer in hot water for a few
minutes, and then fry him in butter.   But he is sweeter
treated *à la juive*.

There can be no doubt but that bream "mud" in very
cold weather, for I have known bottom-fyers throw them
out of the dikes in their dydles.

Though they seldom exceed more than four pounds in
weight, still numerous are the Broadsmen's legends about
them.   "I never knowed a bream 'ter die of old age," says
one.   "There's bream over a hundred year old in this 'ere
broad—ay, old horned bream, with corals all a-hanging ter
'em," says another.   "That there be," confirms another.
"Why, I ha' caught a pike in these 'ere waters without a
tooth in his head," adds a third.   And so on, and so on.

ON A NORFOLK RIVER-SIDE.

CHAPTER XI

EELS

THE life-history of the eel is shrouded in mystery. Even specialists have not solved the problem, nor do they seem much wiser than the eel-catchers of the Broadland. The contradictory remarks, the legends, the different theories, and mysterious habits that surround this fish would fill a volume. Nor can I add anything to clear up the eel-mystery. Indeed, every eel-student's theory or knowledge only renders confusion worse confounded. Some declare there are nine kinds of eels, others six kinds, others four kinds, others only two kinds—which is the theory I am inclined to. Two eels seem to me to differ altogether in habit and custom—the sharp-nosed or silver-bellied eel, and the broad-nosed eel. But let me tell how many the Broadsmen distinguish of other kinds, which may be varieties, or merely sexual or age differentiations.

Some Broadsmen describe—

1. Stream eel, with a frog mouth, that runs in August and September.

2. Common swim eel, running to nine or ten pounds in weight. They aver that neither of these eels can be taken on a hook, and that both have a mark down the sides of their bellies. They are said, too, to possess larger eyes than any other eels, and to "fly" three times as quickly.

3. Silver-bellied eel, which is said to be similar to a stream eel, but never runs above five or six pounds, and has smaller eyes than the stream eel, a silver belly, and black back. He will take a hook.

4. Glut, with a big head, running to two feet in length,

whose average weight is a quarter of a pound. He is said to
have wonderful fins—to be all fins up to his head—and to have
a bigger mouth than the second glut. And they say, " If you
get a drop of grease out of him, it is worth a sovereign."

5. Another variety of glut, which, they say, runs larger—to
two or three pounds—will **take a hook**, but is an extremely
dry fish.

6. The grig, said to hang to the weeds, and seldom to
exceed half a pound in weight.

7. A black eel, that does not run nor mud.

8. Congers. Some aver that the conger is nothing but
the " run eel, who tarns ter conger when he get ter sea, and
don't come back **no more**."

It is unwise to be a Philistine, and pooh ! pooh ! these simple
theories of simple minds ; for, as in the case of the beldame's
pharmacopœia, there may be grains of gold in the quartz.

But, judging merely from my own experience, I believe there
are two distinct varieties of fresh-water eels, and only two.

1. The silver-bellied or **sharp-nosed eel**.

2. The broad-nosed eel.

And the points **of difference I note between the two are—**

Sharp-nosed, or Silver-bellied Eel. (Anatomically different from the broad-nosed eel.)	Broad-nosed Eel.
These migrate to sea in autumn.	These do not " run."
These spawn in æstuaries (?).	Spawn in weeds in the broads (?).
More lively by **nature**.	More sluggish by nature.
Far better taste.	Not good to eat.
Keep deeper in water.	Fond of clinging to weeds near surface.
Seldom " mud."	Always " mud " in September.
Take hook greedily.	Do not often take hook.
Travel far.	Do not travel far.
Run to six or seven pounds in size.	Seldom **exceed two pounds**.
" Turned up " by frost.	Not **killed** by frost.
Appear in waters in April after rain, and on a dark night, and **run** till January.	Appear in waters in June, and feed irregularly.
Caught by babbers.	Rarely caught by bab.
Not often got in skim **nets**.	Got by "skimming" in August mostly.

Eels will eat anything fresh—they do not eat carrion. They are fond of baits, such as birds, beasts, frogs, and worms, as the babbers know well. But the water must be thick. If the water is thick, and the weather propitious (S. wind), you may catch as many eels by day as by night.* Eels, as is well known, will travel over river-walls, from the dikes to the rivers, or *vice versâ*, choosing dewy grass to travel upon.

The broad-nosed eel bores into the dike-shores, and has been dug out buried two and a half feet into the shore, coiled up like a snake. I know of one dug out four feet from a dike, in the solid marsh.

The methods of capturing eels in the Broadland are numerous, and some of them wonderful.

1. Netting.†
2. Darting, or picking, or pitching—7 lbs. being the biggest got by darting.
3. Skimming.
4. Babbing.
5. Hooks—night-lines and liggers for big eels that lie in pea-soupy beds of rivers.
6. Eel-baskets.
7. Bow-nets.

"Skimming." The Broadsman skims with a monster landing-net, which he "deeves down" into the weeds of the broad (chiefly in the months of July and August), and pulls up filled with weeds and eels, which he forthwith separates. He prefers to do this in a hot sun, after a rainy night. I have known a stone of eels captured in two hours by this method. They are often taken in the skim-nets just before thunderstorms too. But he must cover his

---

* *Vide* "On English Lagoons," "Wild Life on a Tidal Water," for full description of babbing.

† For full description of eel-sets, netting, babbing, darting, see author's previous works, from "Life and Landscape on the Norfolk Broads" to "On English Lagoons."

trunks with litter or hay during a thunderstorm, or all his eels will go white and slimy, and die. They cannot stand thunder, though it is said they (and tench) will live out of water a week.

On the Waveney they use eel-pods, beginning to put them in at the end of March. These picturesque pods are made of osiers, and measure four feet in length by nine inches in girth at the centre, tapering to four inches in circumference at the ends. They are constructed on the principle of a bow-net, baited with a worm on a rush, and put in on coarse blowy nights.

Special bow-nets, with a small (shale) mesh net, are used for catching eels too. They are baited with roach, and put in close by where the roach and bream are spawning.

Besides the proper eel-sets, the millmen put nets into the dikes below the out-casts. This is done in the month of July, or earlier, if the weather has been warm, and a heavy fall of rain has fallen; for the eels work out of the banks after heavy rains, and begin to run. They will sometimes work out of the ooze, and run in mild days in winter after a heavy rain.

The silver-bellied eels begin to move seaward in September, as is well known, returning the following spring to spawn on the flats of Breydon—at least so say the men of Breydon; whereas I believe the broad-nosed eel spawns in a weed called "lamb's tail," to be found on most of the broads and in the dikes. I have seen this weed full of elvers.

Of course many of the Broadsmen say eels "young;" and one old man says he tells by looking into an eel's mouth whether it has young, and he is fully convinced he has taken sixteen young elvers out of an eel.

But one thing seems certain about the eel: he is capital eating. A nice silver-bellied eel, from half a pound to a pound weight, is the sweetest fish to eat, as a salt-water eel like the Breydon eel is the most delicately flavoured. Whilst living

on the Broads I have cooked them in a *bouillabaisse*, with
vegetables ; fried them (skins on) ; stewed them ; and cooked
them *à la matelote*. The *bouillabaisse* is the best way, I
think, and so prepared they are a delicate dish, fit for any
good sportsman.

AN EEL-CABIN ON GREAT HOVETON BROAD.

CHAPTER XII

THE PERCH

As you walk by the rivers where the kingcups blaze, and the frogs are croaking in the month of April, you will see, perchance, lovely wreaths of perch spawn hanging round the sere amber reed-stalks. And should you unwind one of these spawn-wreaths from the reed and hold it against the azure, you will find it looks more beautiful still, and recalls a hollow tubular bracelet, made of pure glass beads. Some of these lovely necklets lie on the surface of the water, some far below, lying on the weedy depths, most commonly on hairweed. And if you watch the place closely, you will see the exhausted fish at times lying on the bottom after the spawning is over. When they recover a little they return to the deepest holes round about, for the perch loves deep water and a gravel bottom. In these holes they stay feeding ravenously on worms, an they can get them, for they bite furiously and fast after spawning, but are useless for food when caught. Whilst they are getting into condition, the spawn is hatched, about four weeks after it has been fertilised.

The young fry take to the shallow parts of the stream, where the water is warm, and there they stay feeding and growing slowly, always swimming head to stream. In April and May you may see thousands of them swimming against the stream, on the lock floors.

About the end of June the old fish begin to recover from their exhaustion and get fat inside, and afford good sport till the end of September, for they feed avidly throughout the summer, but are fanciful in their meals. As a rule they feed mostly in the early mornings and forenoons, lying dozing at the bottom of their deep holes in the afternoons. Therefore choose a windy morning for perch-fishing, especially when the wind shifts from a quarter it has long been in, and bait with small fry, fresh-water shrimps, or, *faute de mieux*, worms; but remember that perch, like all fish and fowl, hate an east wind. But they prefer small fry (*e.g.*, gudgeons), and you may, if you watch them attentively, see the small fry's scales gleam as he goes down the wide mouth of the hungry perch. A good bait to try is the tail of roach or rudd cut after the manner of a mackerel bait, and I have even known a perch caught with a sounding plummet. For night-lines, worms; or for thick water, young eels are good. You may often in warm weather see a shoal of perch together; and cast your bait just over them, when they will look at it languidly, until suddenly one will dart at it and take it. Others may follow, but perch-fishing is very uncertain sport. In the summer they work about in shoals from one place to another; and you may at times see a perch with bristling spines fly at a spoon-bait or fish, and take it greedily down. But the deeper the water the better they like it, for they are not a mud fish. In winter they are more rarely seen, for they lie in the deepest water for warmth, or in holes under the banks; but they feed in winter as well as summer. I have known them caught at Christmas-time. In winter-time, too, you may see them swimming under the ice, away from Jack, who will take them as soon as anything.

The largest perch I have caught was a three-pound fish—a full fish—but it gave no sport, coming up sluggishly. I was casting for pike with a spoon-bait. There is a good specimen at Geldeston Lock that weighed 5 lbs. 2 oz., and was caught below the weir with a spoon. Barber, the celebrated fish-taker, told me eight pounds and a quarter was the largest perch he ever saw taken in the district.

Small perch make excellent *frittura*, and a good large perch makes a capital dish, as does its roe, nicely salted and fried in oil.

CHAPTER XIII

THE PIKE

THE earliest fish to spawn is the pike, the female beginning to cast her spawn on the first of March. If the weather be open and warm at this season, she will spawn in the dikes and fleet water (about one foot deep) on the edges of the broad, casting her sand-like spawn on the weeds. Should the weather be coarse and cold, however, she will spawn in deep water, where it is warmer; and there you find the empty fish as proof. The spawning period lasts for three weeks, and at this season the needy marshman shoots, nets, and darts them, poor in condition though they be. I have known one pike weighing twenty-nine pounds to have been shot, and another weighing eighteen pounds taken with a dart, both whilst roudding. They will migrate miles to spawn, and prefer the reed-beds if the weather be windy and warm. They get two or three together when spawning, and rub their sides together, or rub against alder-roots. When three be together, two often hold their heads one way, and the middle one his head the reverse way. When the spawning is over they draw off, leaving the spawn on the hair-weed, and return to their particular waters, where for a short period they feed ravenously on their neighbours.

In June they begin to lay on internal fat; and if you examine the big fish " darted " during the summer months, you will hardly ever find any food in their stomachs; and I am inclined to think big fish do not feed at all during the summer months, though they, too, feed just after spawning, till the middle of May perhaps. Those seen feeding in the

summer months are chiefly young fish. The big fish are
not easy to see in the water either during June or July ;
for when the tench begin to spawn, they work down in two
feet of water, and entirely cover themselves over with weed,
occasionally leaving a bit of tail showing—a gleaming patch
that the darter soon espies and strikes at ; the darter choosing
a still close day—not too bright, and when the water is tea-
coloured. The big fish that the dart grips, seem to lie lazily
in their weedy lairs, dozing the summer away, and getting
covered with parasites or " suckers," as the Broadsmen call
them, and mayhap developing those tape-worms they have.

When the autumn gales stir the broads, they grow live-
lier, and go to deep water, where it is warmer and stiller.
There they, too, begin to feed, and never return to shallow
water, unless it be to follow their prey ; for when they are
hunting, with their fins erect and eyes glaring, they will work
through a reed-bed just like a tiger after its prey, taking
their fish crosswise, no matter its size, turning it afterwards,
and swallowing it head first. They often go from their haunts
to feeding-grounds at night, and return at daybreak. An
old fisherman assured me he once saw a pike take a great
old bream that was coming end on, and it choked him, so
that both died. In *On English Lagoons* I recorded a
similar case.

Pike will take any fish when hard pressed—tench, perch,
eels, bream, roach, rudd, or smelts, as well as voles, and
in the dikes young birds, for they seldom take fledglings
unless in confined waters where fish is scarce. On the
open waters, when there is plenty of bait, they do not touch
them, though sometimes they take a duckling.

The pike is a good jumper, like the salmon; and at
the breeding season, if you leave him in a trunk (open),
he will jump out. They will jump out of a pond, too—
ligger and all at times—especially if the water be warm.
He is very strong, and woe to the trunk that is placed
under water with a big pike in it. Pike will jump out of

the water, too, if a gun be fired over the surface on a still
warm day.

Just after a winter's daybreak is the best time to catch
a pike. He is on the feed then, all his fins working, looking
out for prey. During the day he **does** not feed much, but
takes his **dinner in the evening.** During **the** winter he
seldom **feeds at night.** Sometimes during the winter, **when**
the river is laid, he will get into the reed-beds after fry ; **and**
if a good **quanter** goes in after him, he will soon "muddle"
him up ; for he easily loses his head, and can be caught.
Pike soon give up, too, if chased under the ice ; but **they**
will not be "muddled" into a bow-net, going round
and round it, but always avoiding it. Still, they will go
into bow-nets **of** their **own** accord, in the spawning season
especially.

Many "fisherman stories" have been told of the enormous
pikes to be caught in the broads ; but thirty-two pounds odd
is the largest I can hear of, though I know one of these
thirty-two pounders was shown at a large exhibition, and
labelled forty-two pounds. Barber assures me he never got
a pike weighing thirty-three pounds ; **nor** have I ever yet
met any one **but an amateur** fisherman who has caught **a**
pike **in the broads over that** weight. Such fact **may** have
been, however.

Pike do not like salt water ; some of the old Broadsmen say
"it tarns them blind," but that " I must **leave."** Pike are
best fished for in windy weather (easterly winds excepted).
An especially favourable season is a northerly wind suddenly
coming after the weathercock has stood in the west or
south-west for a long period. They are all on the feed then,
and eat ravenously, especially if there be a good breeze, or
even if it be drizzling, without the good breeze. A south-
easterly breeze too is favourable, but if the wind back to the
east you will not get fish.

The commonest bait **used on the** broads is a small roach,
which is **got with a cast-net. The caster** stands in his

punt with his circular net, weighted with eighteen pounds of lead, on his left shoulder and in his right hand, his keen eye fixed on any movement in the shallows under a lee, where the bait are feeding. Presently he detects a pike strike at a bait under a reed-bed, and he directs the sculler towards the place, when the net is cast spreading into a circle as it falls, describing a circle as it strikes the water, the weighted circumference finally describing a cone in the water and closing on the sharp-nosed roach. The liggers are then baited with these roach, and started on the windward side of the broad, and they go "wallering" across, driven by the wind, pricking to windward when taken.* But small perch, and pike, and golden carp are the deadliest baits for liggers.

There is no doubt that the colour of the pike varies according to the water he inhabits and the food he feeds upon; but the female can always be distinguished, for the male is spotted and lighter in colour than the female, which is darker.

Some old Broadsmen aver that pike and other fish go into gravelly shallows just before spawning, to clean the slime off themselves, which comes again in autumn, even collecting over their eyes; but of these things I have no experience.

When the smelts come up the rivers in the spring, the pike eat them, and are very fond of them. One old lock-keeper assures me you cannot open a pike at that season without finding a smelt inside of him. "You may see them inter lock with the smelts." Smelts are therefore a good bait.

Shrimps too are seen in the locks—that is, the pike that eats the smelt that eats the shrimp are seen altogether—all in the locks; and a dainty morsel must an eleven-inch smelt be for a good-sized pike, though the keeper has caught a smelt a foot long; and when we know that fifty score of smelts have been caught in a lock (though twenty score are rarely caught now-a-days), we realise that "Jack" must have a good time.

* *Vide* "On English Lagoons."

An experimental Broadsman told me he once got some pike spawn from some alder roots by a reed-bush, and took it to a dike where reed grew, and the spawn couldn't get away, and that in three weeks they were an inch long.

A curious fact as regards pike is their taking up stations. Should a big pike be caught in a certain part of a broad or · mere, another (the next in size in the district) will take his place; and if he be caught, another takes his place, and so on.

On the tidal rivers the pike go up the dikes on the floods to get out of the way of the salts.

For the table, a winter pike, weighing six or seven pounds, is the best; and for a big pike's food, a half-pound pike is the best bait. All fresh-water fish are best to eat in the coldest months of the year—November to March—whilst many, as bream and roach, are unfit for food in the summer months; they get soft and muddy, whereas in cold weather their flesh is firm and full of flavour.

And the best way I know of cooking a pike is to clean him, being careful to scrape his backbone as clean as a hound's tooth; then place him in brine to soak for an hour; then rinse him well in pure water, and hang him up in the open air to drain. If this be done at night he will be ready for breakfast the next morning, when you must cut him into sections; roll them in beaten-up eggs first, and then in flour, and pop the morsels into boiling oil (Lazenby's olive-oil); and when the morsel is a rich golden brown, take it out and let it drain on blotting-paper, then serve, and tell me when you have eaten a better dish.

In winter pike are poached in a curious fashion by the Broadsmen, for they often draw up into the dikes after the bait, because they can catch them easier there. You may see this going on whilst gulls too are poaching small fry that have jumped up on the ice and got frozen in. But let us watch the men-poachers.

The frosts have come early and bound the waters as with

an iron hand. The once warm expanse of broad is now a
sheet of grey ice, bordered by yellow reed-stalks covered
with rime-frosted tassels. A necklace of white, half ice, half
snow crystal, encircles the mere, and tells brightly against
the dark evergreens and curiously-fashioned elm branches
that toss their leafless twigs against a yellowish-grey
snow-laden sky. There is music too as the steel-girt
feet of the skaters turn and twist, shooting forward or
curving gracefully in endless patterns across the field of
unpolished silver. Bright eyes flash and red cheeks glow
in the dead landscape, and faces turn skywards for a
moment as a wedge-shaped flock of golden plover dart
across the ice, their heads well thrust back into their necks,
bearing a fanciful resemblance to one of the dark muffs into
which girls' slender hands are thrust as far as the gloved and
supple wrists of the wearers. It is a sane and merry sight.
Only the fittest are out there—the old and weak of man-
kind are nestling round the fires at home, whilst the old
and weak birds are lying dead under the hedgerows, stricken
by the rude though invigorating hand of the northern blast
that sweeps across the ice from N.N.E., catching up the girls'
dresses, showing neat ankles, rattling the twisted papers,
dropped by the smokers, that sail like ice ships with a
crackling noise across the ice into some reed bights, and
shaking the powdery feathers from the reed tassels. Gulls
flash against the heavy sky, seeking their quarry buried in
the ice; sometimes they hesitate, hover for a moment, and
fly on. Mayhap they have seen through the ice by yon
reed-bed the sharp snout and greenish shadow of a pike
lying with slowly oscillating fins on the alert for food, or
have their sharp eyes detected the frozen body of a roach
embalmed in the cold crystal? Any way it is hoping against
hope for them, for the fish are securely lodged in their glassy
prisons. Come snow or wind, the pike are in a tranquil
world that rings only with the vibration of the skaters' feet.
Perhaps that is why the pike near the reeds—for all were not

in deep water—started off with a swift sweeping of his fins, as if followed; then swirled in his gloomy prison ere he looked up through the frosted windows with round anxious eyes.

Leaving the skaters, I stole off in search of wilder scenes that I knew were to be found in secluded bays of reed, for as in the city often the most deserted spots are nearest the most crowded thoroughfares, and the garrotter grips his victim on the edge of the roaring lusty crowd of life. I passed a "wake"—or open space in the ice—where the swans were swimming like sentries on duty; two coots were also there enjoying the restricted open water, but they saw me coming and flew off. The swans did not hiss as I passed; they seemed ashamed to express their wonted ingratitude for the care of man. Grebes too were sitting awkwardly on the ice, and water-hens and rails were feeding at the shore rills.

Going along the ice, I suddenly turned into a deserted bay bordered by gladen. It was very "sheer," and I could see the brown bottom a foot beneath me. "Wind-frost ice" was this, sheer and smooth as glass. Before me two peasants armed with clubs were walking along slowly, looking with keen eyes through the ice upon the bottom. I lay down in the rushes and watched; a crime was imminent. The figures walked slowly along—two black masses against the grey ice. Presently one stopped and signalled eagerly to his companion, who was a few yards ahead of him, and they both stopped dead, as if frozen by an instantaneous increase of cold. The man nearest to me raised his club quickly and struck the ice a heavy blow, a sharp note sounding through the still wintry air—a cruel blow. Lifting his club once more, he peered eagerly through the ice-window made by his club and starred with a thousand cracks, said something quickly to his companion in an undertone, and drawing a "shutting knife" from his pocket, dropped quickly upon his knees, and began to dig into the hard ice with might and main. The frosted splinters flew about him, whilst his companion, who had come up, looked eagerly into the hole. The murderer worked with devilish haste, never stopping to look round until a hole was made through which the water gurgled up, staining the frosted ice, killing the brilliant rainbows that flashed in the cracks that had been made by the stabs of his old knife. He stopped for a moment, turned up his coat and guernsey sleeve, and thrust a brown and weather-stained arm into the water and triumphantly drew forth a pike about four pounds in weight. With the club he rapped the victim on the head, placed it hurriedly in a large inner pocket, and with his accomplice went off with the quick soft steps of the criminal.

And many is the pike that has been dazed in winter by such ice-blows, and seized through those rough-hewn holes. *Mais il faut manger. Que voulez-vous?*

"NATIVES" FISHING ON GREAT HOVETON BROAD.

CHAPTER XIV

ROACH

THE roach is a deep-water fish in winter, as the perch and pike know well. But in the most severe weather, like most bottom-feeders, they will draw under the "segs" or floating hovers that the ice has prized from the bottoms. In winter, when the weather is open, they keep to their haunts, and do not go far away—in this respect resembling swans. Nor do they confine themselves to the broads, but are common in the tidal waters; indeed, for eating, a good "salt-water roach" is far and away the best.

The roach, like most fish, feeds mostly at night (except, perhaps, in the depth of winter); so that you may know holes that are full of bream and roach by day, yet at night they will be empty; for they go forth to their feeding-grounds, returning at daybreak, or later. Night, therefore, is the time chosen for netting them; and for the same reason, an angler stands a poor chance by night. Pike "flight" at night, too, often going to the dikes, as do gudgeon, though they will not go near a salt. In fact, fish flight as do birds, and for much the same reasons. By day, then, is the best time for angling; and choose a roach swim in an eddy or tidal water; and let the wind be a south-westerly, blowing with a gentle breeze, for no fish will bite on a perfectly still warm day; and the time to catch them to eat is two or three weeks before they cast their spawn, when they are full; and the way to cook them is *à la juive*, as directed for pike.

At the end of May the roach shoals collect, a few days before the bream, on the shallows, on broad or river edges

on the hair-weed, the males appearing first. They are to
be seen sailing about, their fins out of water, looking for the
spawn, for the males fill first. But the females soon begin
to cast their spawn, and then there is a commotion, the water
being black and troubled with them ; next small roach jumping, and big males' fins cutting the water, all portents that
spawning or roudding will soon begin.

A busy time is the spawning period; for once they begin,
they go straight ahead and finish, like a pike, a very severe
wave of windy cold merely delaying the process for a day
or so. At this season all manner of fish collect (and some
birds, the swan amongst them) round them, and feed on
the spawn and fish—more especially eels and pike. A small
pike will fill himself with roach and bream spawn ; for with
them it is "get all you can whilst it lasts," for in three days
the spawning is over. And you may, an you miss the great
festival, and arrive upon the scene too late, see where the
spawning has been, for the weeds are all pulled about and
broken off; for the spawning roach get the weeds round
them, and break them off, as do the bream. Up some of the
narrow rivers the shoals have made the water so black that
the river seemed full of fish, and wherries seemed to be
sailing through a sea of roudding bream and roach. It seems
that during the roudding period all fish must rub against
something, either each other's bodies, or roots or boulders,
and so the bow-net comes in most useful at this period.

When the great spring festival of the spawning is over,
they return in shoals to their haunts in the deep water.
And this accounts for the fact that on some waters no roach
are to be seen except during the spawning season.

Whilst the spawn is hatching, and the bait growing in
the shallows of the broads and in the shallow dikes, the big
roach are recuperating from their efforts, and are not worth
eating till the end of July ; indeed, between the spawning
and the day shooting comes in, roach are not in condition,
and therefore not worth catching, unless you be a vain

commercialist, whose sole idea of life is number and advertisement.

The roach's chief food is weed, which they get from the bottom; but gentles and worms are the surest baits. On the tidal rivers, Mr. Grimsell of Reedham, an old Lea roach fisherman, and an expert, tells me he always ledgers in the tide-way, using eighteen yards of fine gut-line, one foot of the finest gimp, two feet of the finest gut, a lead shot, and a Pennel hook. He fishes on the top of the flood. The swim is ground-baited with balls made of boiled wheat and rice, some uncooked wheat, and the ball is filled with bran in the centre. Besides these balls, he throws uncooked wheat against the tide, for it settles quickly; and so he *keeps his fish together*, which, he says, is the great thing in roach fishing. If they do not bite well, he throws them a few grains of cooked wheat. Mr. Grimsell uses an eighteen-foot split cane rod, and *ties* his line to his rod about three-quarters of the way down the rod, thus fishing with a tight line. Mr. Grimsell's heaviest roach weighed one pound thirteen and a half ounces, a full fish.

CHAPTER XV

RUDD

THE rudd, or "roud," as he is called in the Broadland, is one of the most beautiful and most useless fishes of the district; he is not worth cooking at any time, and even the neediest Broadsman turns his nose up at the useless "roud." When freshly taken from the water, his blue-bronze back and golden stomach, edged with crimson fins, and dotted with red-rimmed eyes, make him a handsome object. Each particular fan-shaped scale too is a work of art; so that the fish, on closer examination, appears to be covered with golden and blue-bronze fans. On holding the translucent scale up to the light, you see a series of concentric markings—a fan of frosted silver opposing the golden seen through an opalescent film. Altogether it is a most lovely and dainty thing, this fish's armour-plate of gold and bronze; and he is well worth the catching, to study his wondrous beauty.

During the winter months he too takes to deep water, in shoals—some large fish, others small fry. He chiefly feeds at this season on weeds, biting the tops off the "muzzle-weed" as cleanly as a rabbit bites off the corn-stalks. But as soon as the spring solstice warms the water, the shoals draw into shallow water, and begin to add flies to their bill of fare. Towards the end of May they get into the shallowest water by the broad edge to spawn, collecting in small shoals. And very busy they are whilst spawning—splashing and rubbing against the roots and hovers, then resting for a quarter of an hour and beginning again—so this strange festival is held. The rudd seems to cast his bream-like spawn in two batches, for they spawn again in the middle of June. Whether these be old or young batches is doubtful.

Soon after spawning they go on the feed, still keeping most of the time to shallow water, though occasionally wandering far afield, whither the flies *seem* to follow them. On a warm day, after a shower of rain, you will find thousands of rudd against the shores of the broad, feeding on " muzzle-weed " and flies.

Though a suspicious fish, often taking food and ejecting it again, they will rise to a fly; indeed, that is the only way to catch the large ones, except, of course, by netting. If you want to go fly-fishing for rudd in July and August, choose a time when a gentle breeze is rippling the water, and then you may see them rise all over a shallow broad. The osprey knows that too, and rudd is his favourite fish; for as the rudd keeps near the surface for the flies, the osprey has no difficulty in catching him.

Anglers attack the rudd with gentles, worms, and paste, but paste is one of the best baits; indeed, he is a better " paste-fish " than a roach. He bites best after a rain-shower on a hot summer's evening. But he seldom scales more than two pounds, and is, as I have said, useless for food.

CHAPTER XVI

SOME OF THE MORE UNCOMMON FISHES

CARP

Are occasionally caught running to 6 lbs. in weight, and they have been seen roudding at the end of June and beginning of July, when they are most easily caught. They have been seen in Hickling and Barton Broads, and in Irsted shoals, being from all accounts, commonest in Barton Broad. The golden carp is one of the deadliest baits for a pike. An old Broadsman assured me he once caught a golden carp weighing $7\frac{1}{2}$ lbs. in "his broad."

SMELTS

Ascend as far as the locks, of course; and I know a trustworthy man who has caught smelts both at Catfield and Stalham Staithes, but they are not fresh-water fish.

DACE

Are found in the district, some measuring nine and ten inches in length, February being the best month to catch them. They generally angle for them with paste below the locks, weirs being good places for them. I have caught ten or twelve in one day. The lock-keepers get a number of these fish at the smelting season as well. Some think them better than roach for pike-bait. They get into the mill-dikes too, and work against the stream; but they seem to disappear as the roach and bream collect for spawning.

GUDGEON

Have grown rare in the districts best known to me, though in some places they used to be common enough. Their delight is to bask in a sandy shallow on a hot summer's day.

MILLER'S THUMBS

Occur, but are not common. They get them in the smelt nets at the locks.

THE RUFFE,

Or "pout," as the Broadsmen familiarly call him, is common in some of the shoals, and in some of the broads. Anglers do not like him, for they say, "When he come you may as well pull up and go," and they do. Yet he is good eating, though small, eight inches being the longest "pout" I have seen. Pouts will take a worm on a small hook. They prefer a stream and a hard, clear bottom, and, as the Broadsmen say, "they make a face at you when you catch them."

SEA-LAMPERNS

Work up some of the rivers in the autumn, six or seven weeks before Christmas. There are a couple of good specimens caught at Geldeston Lock, now in the keeper's possession; they weighed two pounds and a quarter each in the flesh.

"LAMPERS."

The fresh-water lamprey can be seen in spring hanging to the stones in the same locality. They are sometimes caught in the eel-nets, and especially in snowy water. They have been found with spawn in them on the Waveney.

CHAPTER XVII

TENCH

THIS succulent and lusty fish is the primest table fish of the Broadland; and I shall die happy when I think of the numbers of lusty tench I have eaten, cooked in milk or fried in butter. The old monks were not "all-fool," as they say in Norfolk.

The tench's "coming up" (from the mud) is one of the portents of spring; for though the water may warm in early spring, yet the tench will not come up unless spring have really set in—and they seem to know. So that the amphibians pass the word round joyfully, "The tench be up." And should you be fortunate to get one of the first risers from his wintry sleep—a black-looking fellow he will be, too—you must open his stomach, and you will invariably find little shells therein. Sometimes they come up marked with yellow, so that some old marshmen say they "have been laying in the reeds, and growed a bit, so that mark come." Reed-cutters have dug them out whilst dydling, and found them two feet deep in the mud, head downwards. Some—for they do not mud in shoals—aver those found head downwards were flying from a pike. Those dug from a depth immediately opened their mouths and kept them gaping for ten minutes when thus unceremoniously dragged from their winter beds.

In the spring the smaller fish rise first; they are probably mudded nearer the surface, and feel the genial heat first. They eat best soon after they are up, and get "washed out," and fattened up on "weeds off the bottom," worms, and small mollusca—some of the finest weighing four pounds, some measuring twenty-two inches in length.

Before they begin to cast their greenish spawn, which is early in May—for tench do not spawn in shoals, but spawn here and there in parties from May till July; indeed, all the tench have spawned by the time the wheat is in flower —the Broadsmen are on their track with landing-nets and poles, darts, their fingers (by tickling), and bush-nets; and, later on, once they begin to spawn, with bow-nets.

At this season, on a fine warm April day, the Broadsman goes down to the broad, and sets his bush-net round a reed-bush on the channel side of the reed. The net is several yards long and three feet deep, weighted with leads and floated by corks. The old tench-catcher gets into the thin jungle of reed, sere bolder, and rising lily-leaves—a jungle alive with midges in the still hot air—and "plounces" his quant, driving the fish before him towards the channel and the net. And you may distinguish the fish as they fly startled before the old fisherman in his marsh-boat—the pike fly like lightning through the water, rattling the sere reed-stalks, but do not run so far; the bream go farther than the pike, but move more slowly; whilst the tench run and bubble, going "any way," and not in a straight path, as do the bream and pike. In fleet water, of course, you can see the fish themselves, but in deep water you know where they bring up by the shaking gladen or reed-leaves. A tench does not like too much of this treatment, and if flushed three or four times, he grows sulky, and will not stir till prodded with a pole. Sometimes a stamp of the foot on the boat may do it, however. Perhaps the old fisherman will fish for tench in three ways in the same place. He has his bush-net spread, and he takes his landing-net and drives sulky tench into it with his quant, or else he may bare his arm, and feel for the fish (taking care not to touch its tail), and the fish must not see him. Of course he knows the fish's whereabouts by the moving reed or gladen, for he watches him when he runs, and softly works his low marsh-boat up to the fish. But this method is practised more as a *tour de force* when he can net, or as a method of poaching when netting is not permitted.

By these methods a number of strong, lusty tench can be got in a morning—fish averaging two or two and a half pounds—nor do they take long to capture. But once May is in, and they begin to spawn, the bow-net * is brought into use, being placed in runs in the gladen or reed, or in holes in the dike shores, for a favourite resort of tench at this season is in holes where there is a draught of water. Sometimes flowers are placed in the bow-nets—water ranunculi, calceolaries, and pelargoniums having been used, for it is said these bright-coloured flowers attract the fish. But it is the bow-net that attracts—they like to rub against it, for they do not spawn in shoals, and so have not their comrades to rub against. They choose no particular weed to cast their greenish spawn upon, and seem rather casual, choosing often shallow water in a broken-down gladen-bed or dike. After they have done spawning, they go back to deep water, as do other fish.

In August most of them will take a worm, and old Broadsmen advise you to decapitate the worm, and fish with its tail upwards. Lightning will kill tench in a trunk, and it is the habit with Broadsmen to cover their trunks with litter during a thunderstorm to prevent this.

When the frosts come they mud again, and so run their yearly round.

If a broad grow up too much, the tench will disappear; but where and how they get away, is a mystery—to me at least.

GOLDEN TENCH.

I have never seen a golden tench taken from these waters; but some old Broadsmen scout the idea of golden tench, and say any tench, if put into a clear stream with a gravelly bottom, will turn "golden."

* *Vide* "Life and Landscape on the Norfolk Broads."

CHAPTER XVIII

FROGS AND TOADS

Two frogs and three toads are known to the Broadsmen.

1. Common garden toad, or land toad.

2. Water-toad—a lighter-coloured creature frequenting the dikes.

3. Running toad, with a bright yellow stripe up his back; the frequenter of hedgerows and marshes.

The two frogs are :—

1. **Common** frog of the dikes, varying in size and colour.

2. Small yellow frog, or land frog, found in the harvest-fields; the children delighting to prick him, when he is said to " shriek like a child."

The Garden Toad

Makes a form in the grass during hot weather in which to **shelter** himself; and should you come upon him, he will squat with his bright eyes fixed upon you all the time.

They **are** very fond of beetles, and you may see them in **an** onion-bed sight their quarry (invisible to you) ere they **run** two or three inches and dart upon it.

The Water-Toad

Is found in the breeding season, the early spring, in the warming dike waters, rolling over and over under water, or sitting wrapt in each other's arms (the little male above) under water on some water plants.

The male is darker in colour than the female; and you may

often surprise the couple on the dike-edge, when the female will make for the water, carrying her lover and diving in with him on her back.

As you listen by the dike-side, you may hear the sharp waterhen-like note of the water-toads ; but directly you approach the noise they dive and are gone, for they are very quick to take alarm.

The spawn is found soon after the frogs spawn, and before the running toad spawns. If you go very softly in the middle of April (the height of the spawning season) by the dikes, you may see some amorous pairs with their noses just above water, and others sitting down on the weeds.

A month after the spawning season all is quiet in the dikes, and they are more rarely to be seen, though you often come across them on the walls with their human-like walk—*i.e.*, like a man walking on all fours.

THE RUNNING TOAD,

As I have said, has a yellow stripe down his back, and is commoner in some districts than others. The chief thing in connection with this creature is the rockstaff that a man can quiet the most restive horse with the bone of a running toad, which, it is said, will swim against the stream. Yacht designers and others might well look into this matter.

THE COMMON FROG.

In March the batrachian life of the swamps awakes, and you may any fine morning hear the croaking of the frogs, coming, as it were, from the bowels of the earth. If you go down to the low-lying wet marshes, where grows reedy litter, and search the clear pools, you will presently come across a pool alive with marsh frogs, making the still water bubble in miniature storm. As you approach they vanish, diving and hiding under the stuff, and mayhap a few dark-coloured males eye you from the weedy floor.   Should you leave this pool and go on, you may come across couples collected in bunches that grip each other so fiercely that some are at times suffocated and killed, while others have lacerated armpits, and roll off, dazed and swollen, on to the mass of spawn hanging to the reed-stalks, or lying on the bottom of the weeds—a crystalline blanket, studded with black pilules.   Should you try and separate the couples, you can scarce do it, so closely do they cling.   Later on they become more alert, and will, like the toads, dive into the water as you walk up to them.

These creatures begin their croaking chorus at all hours, according to the weather; but the chorus is loudest in the hottest part of the day, say at three o'clock.   And this croaking is kept up for a month after the love-making season is over.

All through the summer you may come across them, but in autumn they vanish, hiding below the stuff in the dikes and in the mud—the dike-drainers and dike-cutters often turning them out with the old litter in bottom fying, or dig them out when trimming the shores in the winter-time. In hard winters they get frozen to death, and "turn up" with dead eels when the thaw sets in.

The Land Frog

Frequents grassy places, and often lies with his feet over his eyes.   He too, like the toad, will make a form in the grass in hot weather.   He is fond of worms; and it is amusing to capture one and starve him for a bit, and then let him out on to the grass, placing a clean, lively worm in his path.   He sits there breathing hard, then his quick eyes catch sight of the worm, and he gives a hop and darts upon it, putting it into his mouth, and brushing the wriggling creature with his fore-feet as it descends into his throat.

After rain you are sure to see these frogs hunting for insects and flies.

Their shrieks when cut in two, as often happens by the mowers' scythes, are heartrending, and resemble those of a child, as do they when killed by a rat or shrew mouse; for the little shrew mouse can and does kill frogs and toads.

WOOD-SORREL.

CHAPTER XIX

VIPERS

You will find two kinds of vipers on the marshes—the common viper, and a smaller "red viper." Both frequent the "walls," and are not uncommon.

As you walk along the desolate country early in the year (as early as Good Friday), you may come across sluggish vipers sunning themselves. At this season they are fattest, for they "lay up" in holes in the bank in the winter; but in the spring, after a shower of rain, when the sun bursts forth, transfiguring the landscape, they come forth to enjoy the day—both the larger and blacker female, and the lighter and smaller male. Also, the smaller red viper are, upon such days, to be seen coiled up on the walls or marshes; in short, in any position where they can lie dry.

There they catch bumble-bees, beetles, frogs, and mice. An old marshman told me he once saw a viper "setting with his chaps open, and a big marsh-bee come and buzzed round and round his mouth, till that went straight in." They will lie in the dikes, too, with their open mouths just above water, and trap the water-newts (swifts) "squiggling about jest like an old eel."

Should you be quick enough to catch the tail of one as he glides into his hole, and pull upon it, you will find it a very difficult thing to pull him forth, as you will a rabbit or a rat in the clutches of a ferret. The old marshmen say they "set out their scales to hold their selves," but I cannot tell whether this be so.

In the summer-time they are fond of lacing themselves

through a clump of grass, or lying in clumps of sallow or bramble on the walls, or on a beaten path. Some marsh-men aver the young will run down their mothers' throats, and others swear they "young" like an eel.

Their bites, though not fatal, are severe. I have known of a small dog killed by a viper's sting; and I know a strong, healthy marsh-mower was once carried home as if dead after being stung by one of these hateful creatures.

I have heard experienced marshmen say that "floods kill the wipers," but I know nothing of that.

There is a curious rockstaff in the marshlands that a viper's slough will draw thorns from your flesh, whilst "wiper's oil" is a reputed specific for "screwmatics."

GREAT HOVETON BROAD.

Printed by BALLANTYNE, HANSON & CO.
Edinburgh and London